RESEARCH AND PERSPECTIVES IN ALZHEIMER'S DISEASE

Fondation Ipsen

Y. Christen P. Churchland (Eds.)

Neurophilosophy and Alzheimer's Disease

Springer-Verlag
Berlin Heidelberg New York
London Paris Tokyo
Hong Kong Barcelona
Budapest

Christen, Yves, Ph.D.
Fondation IPSEN pour la Recherche
Thérapeutique
30, rue Cambronne
F-75737 Paris CEDEX

Churchland, Patricia S.
Department of Philosophy
University of California
La Jolla, CA 92093
USA

ISBN 978-3-642-46761-5 ISBN 978-3-642-46759-2 (eBook)
DOI 10.1007/978-3-642-46759-2

Softcover reprint of the hardcover 1st edition 1992

Reproduction of the figures: G. Dreher GmbH, 7000 Stuttgart
Typesetting: Mitterweger GmbH, 6831 Plankstadt

27/3145/543210 – Printed on acid-free paper

Preface

Any mention of the relationship, still poorly understood, between body (or brain) and mind invariably invokes the name of Descartes, who is often thought of as the father of modern philosophy and perhaps of neurophilosophy. Although a native of the heart of France (the region around Tours), René Descartes travelled widely, as everyone knows, especially to Holland and Sweden. It should come as no surprise, that the Congress of Neurophilosophy and Alzheimer's Disease was the first in the series of Fondation Ipsen Colloques Médecine et Recherche to be held outside France.

The meeting was held in San Diego (California) on January 11, 1991. This venue was chosen for a number of reasons. The University of California San Diego is without doubt one of the most dynamic universities today. A good number of friends of the Fondation Ipsen who have taken part as speakers in previous conferences are based there. Patricia Churchland, whose publications have helped "launch" the term "neurophilosophy", also teaches there. The choice of this particular venue gave us the welcome opportunity of benefiting directly during the conference from the participation of many eminent (including some Nobel Prize-winning) scientists, including biochemists, neuroscientists and "alzheimerologist", psychologists, cognitive science specialists and philosophers.

In this volume we have published the texts of speeches presented on this occasion. This book forms part of our series dealing with Alzheimer's disease, and follows a list of titles which are perhaps less astonishing in academic terms: *Immunology and Alzheimer's Disease, Genetics and Alzheimer's Disease, Neuronal Grafting and Alzheimer's Disease, Biological Markers of Alzheimer's Disease, Imaging, Cerebral Topography and Alzheimer's Disease,* and *Growth Factors and Alzheimer's Disease.* The next volume in the series, *Heterogeneity of Alzheimer's Disease,* will be published after a conference in Marseille, France, on April 6, 1992. The central idea underlying the choice of the theme for the present volume is the following. Alzheimer's disease could be a good model for the study of typical neurophilosophical problems such as the mind–brain relationship and also, more particularly, as a means to understanding different aspects of human consciousness and thinking. In effect, Alzheimer's disease presents a situation in which both modification of the cognitive faculties and organic changes, cellular and even molecular, may be observed simultaneously. Since this disease is currently at the centre of a considerable number of research programmes in exellent laboratories, a large volume of data is available. Thus,

Descartes and the mind-brain relationship. This famous picture was published in *Traité de l'Homme* in 1664. It illustrates the functioning of the human machine as a result of the discharge of the "esprits animaux" from the pineal gland of the brain

much work of little apparent philosphical interest takes on great significance when viewed in this light.

The dominant trend in "alzheimerology" is currently amyloid protein. Despite some deeply interesting work in the field of molecular biology, the role of this protein in the disease is still not clear. It is not known with certainty whether it is the cause of the disease (many specialists think so), simply a consequence of the disease, or possibly even a defensive reaction of the host organism against an unidentified pathogenic process. It is clear, then, that the central question of the most fundamental research deals with a problem of a philosophical nature, namely the search for a *cause,* and more especially the cause of major cognitive modification. Francis Crick arrounced somewhat provocatively in his memoirs the advent of molecular psychology[1]; one may

[1] Crick F (1988) What mad pursuit: a personal view of scientific discovery. Basic, New York

wonder whether this type of work does not, in fact, constitute the premise of such a statement. It is reasonable to expect that these data should take on a more general significance beyond the confines of research into Alzheimer's disease. In science, the pathological is often used to understand the normal. The study of illness has led to improvements in understanding physiology, and the study of mutation has greatly contributed to the advancement of genetics. It is interesting to note that Emil Kraepelin, a pioneer of neurology and Alois Alzheimer's teacher (and the first to use the expression "Alzheimer's disease"), wrote:

> An acquaintance with these disorders [of mind] unlocks a mine of discoveries with a direct bearing on everything connected with the life of the mind. The study of dementia does not simply uncover a body of general laws; it opens up before us a wealth of profound insights into the history of the development of the human mind, both of the individual and of the entire human race[2].

Commenting on this quotation, Marc Jeannerod wonders whether "the range of psychiatric disorders might not be a simple inverted or negative image of a psychic repertory containing the rules with which we construct our behaviour and our personality"[3].

It is therefore most useful to have models, preferably many models, in which the breakdown of the mind–brain relationship illuminates that which is normally transparent, namely normality. Reflection on such models is evidently already underway. Studies on the phenomenon of the split brain have already provided a celebrated example[4]. On a purely theoretical level, the reflections of Daniel Dennett[5] on experiments in brain transfer provide another model in their own way.

The interest in the Alzheimer model no doubt stems from the fact that it is related to a particularly common situation, which is by no means the case for the other two examples given. It is certainly not the aim of either the conference in San Diego or the present proceedings to pronounce the last word on a debate which is certain to continue for a long time to come. Our more humble intention was simply to help introduce the neurophilosophical debate into an area of research where it is more often molecules that are manipulated rather than philosophical concepts.

I would like to thank all those who contributed to the conference in San Diego and to the present publication. My thanks are due in particular to Floyd Bloom (La Jolla), Hanna Damasio (Iowa City), Roger Guillemin (La Jolla) and Larry Squire (La Jolla), who agreed to chair the sessions; Jacqueline Mervaillie, who took care of organisation down to the very last detail; and Mary Lynn Gage for her editorial assistance.

Finally, I would like to acknowledge my very speical debt to the late Jean-Louis Signoret (1933–1991), professor at the Hôpital Pitié-Salpêtrière, with whom I frequently discussed both the subject of this book and the conference before it was actually organised. Signoret presented a paper in San

---

[2] Kraepelin E (1984) Introduction à la psychiatrie clinique. Navarin, Paris (1st edn 1900)
[3] Jeannerod M (1991) Esprit, où es-tu? Jacob, Paris
[4] Gazzaniga M (1985) The social brain. Basic, New York
[5] Dennett DC (1978) Brainstorms: philosophical essays on mind and psychology. Bradford, Montgomery

Diego which is included in this work, and he came back from the conference filled with enthusiasm. Alas, he was to take his leave of us a few months later, the victim of a heart attack. I would like to dedicate the present work to his memory as a token of my friendship and my gratitude for the help he gave, from the very beginning, to all the initiatives of the Fondation Ipsen concernced with Alzheimer's disease.

Yves Christen

Contents

Contributors

Changeux, J.-P.
UA CNRS D1284, Molecular Neurobiology, Institut Pasteur,
25, rue du Dr Roux,
75724 Paris CEDEX 15, France

Churchland, P. M.
Department of Philosophy, University of California, La Jolla,
CA 92093, USA

Churchland, P. S.
Department of Philosophy, University of California, La Jolla,
CA 92093, USA

Ciaravino, J.
Department of Psychiatry, NYU Medical Center, 550 First Avenue,
New York, NY 10016, USA

Convit, A.
Nathan S. Kline Psychiatric Research Institute, Orangesburg,
NY 10962, USA

Damasio, A. R.
Department of Neurology, Division of Behavioral Neurology & Cognitive
Neuroscience, University of Iowa College of Medicine, Iowa City,
IA 52242, USA

Damasio, H.
Department of Neurology, Division of Behavioral Neurology & Cognitive
Neuroscience, University of Iowa College of Medicine, Iowa City,
IA 52242, USA

Dehaene, S.
UA CNRS D1284, Molecular Neurobiology, Institut Pasteur,
25, rue du Dr Roux, 75724 Paris CEDEX 15, France

Dennett, D. C.
Center for Cognitive Studies, Tufts University, Medford, MA 02155, USA

Dror, I. E.
William James Hall, 33 Kirkland Street, Cambridge, MA 02138, USA

Ferris, S. H.
Department of Psychiatry, NYU Medical Center, 550 First Avenue, New York, NY 10016, USA

Flicker, C.
Department of Psychiatry, NYU Medical Center, 550 First Avenue, New York, NY 10016, USA

Gazzaniga, M. S.
Dartmouth Medical School, Hanover, NH 03756, USA

George, A. E.
Department of Radiology, NYU Medical Center, 550 First Avenue, New York, NY 10016, USA

Golomb, J.
Departments of Psychiatry and Neurology, NYU Medical Center, 550 First Avenue, New York, NY 10016, USA

Kluger, A.
Departments of Psychiatry, NYU Medical Center, 550 First Avenue, New York, NY 10016, USA

Kosslyn, S. M.
William James Hall, 33 Kirkland Street, Cambridge, MA 02138, USA

de Leon, M. J.
Department of Psychiatry, NYU Medical Center, 550 First Avenue, New York, NY 10016, USA

LeSueur, L. L.
Dartmouth Medical School, Hanover, NH 03756, USA

McRae, T.
Department of Medicine, NYU Medical Center, 550 First Avenue, New York, NY 10016, USA

Narkiewicz, O.
Institute for Basic Research, Staten Island, NY 10314, USA

Rapoport, S. I.
Laboratory of Neurosciences, National Institute on Aging, National Institutes of Health, Bethesda, MD 20892, USA

Reisberg, B.
Department of Psychiatry, NYU Medical Center, 550 First Avenue, New York, NY 10016, USA

Rubert, E.
Yerkes Regional Primate Research Center, Emory University GA 30322, USA

Rusinek, H.
Department of Radiology, NYU Medical Center, 550 First Avenue, New York, NY 10016, USA

Savage-Rumbaugh, E. S.
Department of Biology, Georgia State University and Yerkes Regional Primate Research Center, Emory University, GA 30322, USA

Signoret, J.-L.
Hôpital de la Salpêtrière, Clinique Maladies du Système Nerveux, 47 Bd de l'Hôpital, 75013 Paris, France

Smith, G.
Departments of Psychiatry and Radiology, NYU Medical Center, 550 First Avenue, New York, NY 10016, USA

Terry, R. D.
Department of Neurosciences, University of California, La Jolla, CA 92138, USA

Tsui, W.
Department of Psychiatry, NYU Medical Center, 550 First Avenue, New York, NY 10016, USA

Wegiel, J.
Institute for Basic Research, Staten Island, NY 10314, USA

Wisniewski, H.
Institute for Basic Research, Staten Island, NY 10314, USA

Wolf, A. P.
Department of Chemistry, Brookhaven National Laboratory, Upton, NY 11973, USA

Introduction: Neurophilosophy and Alzheimer's Disease

P. S. Churchland

A topic which is at the very center of philosophy is epistemology – the nature of knowledge. Systematic inquiry began with Plato, who was deeply puzzled about how humans learned and remembered anything, including, for example, facts, faces, theories, and mathematical proofs, as well as skills such as carpentry or statecraft. In focusing on mathematics, Plato found the problem of learning so intractable that he concluded that mathematical knowledge was actually not learned at all.

Plato realized that the deepest problems were not confined to mathematics, but centered on the epistemological feats of *abstraction and generalization.* How, he wondered, can we learn to recognize many different entities as all examples of cats, or pots, or storms? Despairing of an answer, he argued that general knowledge was innate, and genuine learning was only illusory. Learning seemed to occur only because at birth specific knowledge was forgotten and required experience and instruction to be recovered. Ostensible learning was in fact recollection. Plato's theory had a number of grave difficulties. First, the mechanisms whereby information is stored and recovered are no less puzzling and problematic than the nature of how something might be learned in the first place. Secondly, it was troublesome because it failed to address either the source of the original knowledge or the vehicle for prenatal knowledge. The awkwardness of Plato's solution notwithstanding, the questions he posed and the alternatives he considered set the basic agenda for inquiry for about two thousand years.

Advances in empirical understanding have allowed us to approach the basic problems concerning learning and memory with more specific questions and to discern at least the form some answers will take. Evolutionary biology provides the general framework to explain how certain kinds of knowledge could come to be built into the brains of a species. The discovery of the structure of deoxyribonucleic acid provides the basis for understanding the vehicle by which certain aspects of knowledge can be specified in the brains of individuals. Developmental psychology investigates the scope and limits of innate capacities in humans and other animals, and the distinct contribution of experience to an individual's knowlege of the world. Research in experimental psychology that focuses on learning and memory continues to reveal aspects of the learning process such as time constants, the role of attention and awareness, the role of salience and context, and the durability of some kinds of knowledge relative to others. Neuropsychology, in analyzing the residual capacities in brain-damaged humans, provides unique links between behavior and brain structures.

Until the recent flowering of neuroscience, particularly within the last three decades, ideas about *mechanism* – about how exactly a physical organ like the brain could learn and apply the information in appropriate conditions – were necessarily vague and speculative. Indeed, until quite recently, a not uncommon opinion outside of neuroscience has been that it is not actually the physical brain that learns and remembers; it is the nonphysical, Cartesian soul. The shift from purely speculative theories to empirically grounded hypotheses about both the functional organization and the mechanisms of learning and memory implies a *naturalizing* of epistemology; that is, the acquisition and application of knowledge are seen as part of the natural world, subject to empirical investigation and understanding, just like any other natural phenomenon. Although naturalism goes against the grain of many philosophers who assume that pure reason alone, independently of empirical science, is the sole avenue to genuine understanding of these issues, the remarkable progress in empirical inquiry favors philosophy's merging with the scientific mainstream rather than perpetuating its scientific isolation.

In nervous systems there is both large-scale and small-scale organization, and different operations take place on different levels. Organization ranges from the level of the molecule, to the single neuron, to small networks, large networks, areas, systems, and finally, the whole central nervous system. One sort of account will explain how signals are integrated in dendrites; a distinct account will explain the interaction of neurons in a network or the interaction of networks in a system. A model that captures the salient features of learning in networks will have a different face from a model that describes the properties of a membrane channel such as the *N*-methyl-*D*-aspartate receptor-ionophore complex. Nevertheless, theories on one level must mesh with theories of levels both higher and lower, because an inconsistency or a lacunum somewhere in the explanatory chain means that some phenomenon has been misunderstood. After all, brains are assemblies of neurons, and something would be seriously amiss if neurons under one description had properties incompatible with the same neurons under another description.

A reduction in science is essentially an explanation of phenomena at one level of organization in terms of the properties and interactions of entities at a sublevel. When it is claimed that the theory of optics, for example, reduces to the theory of electromagnetic radiation, what is meant is that optical phenomena (e.g. refraction) are explainable in terms of the properties of photons. In the context of neuroscience and psychology, we want to understand such matters as visual perception, reasoning, and learning. To seek a reduction, therefore, means seeking explanations of higher level phenomena, such as the recency effect or long-term recollection of events, in terms of lower-level properties. Given the multiple levels of organization between the nervous system as a whole and the individual molecule, the reduction can be expected to proceed step-wise, from one level to the next. The process of discovery, however, requires simultanoues research on all levels of organization, so that any given line of research can profit from results discovered at other levels.

A major puzzle in understanding how a physical system like the brain stores information and then applies appropriately what it stores is this: the physical changes must be local, in many individual neurons. The overall effects, however, is globally coherent. How is global coherence achieved assuming that representations are distributed among many neurons in the brain, especially where some neurons are not in direct contact with others that contribute to the representation? Computer models of neural networks articulate a very general framework for understanding the relation between local changes and global coherence. Activation of one unit by other units, together with a simple rule for updating synaptic strengths, are the basic ingredients of networks that can be trained by exposure to examples of X successfully to recognize new instances of X, where X may be some very complex pattern. In consequence, the feat of abstraction that so puzzled Plato now seems a little more understandable. Exactly how real neural networks learn to recognize patterns is not yet understood, but the demonstration of this capacity in artificial neural networks helps generate working hypotheses about real neural networks (see Churchland and Sejnowski 1991).

In real neurons there are undoubtedly a host of mechanisms for adaptation, unlike artificial neurons which typically have just one parameter of change: namely, the increasing or decreasing of existing synaptic weights. Real neurons are far more complex. They may form new connections and disrupt old ones; new receptor sites may emerge; there may be changes in gene expression resulting in changes in neuronal responses; there may be changes in transmitter released per spike, as well as changes in the volume of transmitter per vesicle. After-hyperpolarization (AHP) may lengthen or shorten, and there may be short-term changes such as post-tetanic potentiation.

The problem of local change and global coherence at the network level is recapitulated at the level of the neuron. Here, a coherent change in a neuron's overall behavior is achieved by means of assorted local physical changes occurring at many different locations all over the neuron. In order for computer modeling to be useful in explaining computation, representation, and modification within the individual neuron, it will be important to simulate real neurons in all their biophysical complexity. It will also be important for neuroscientists to unearth even more detail about membranes, channels, gene expression, and so on, so that the modelers may constrain their models appropriately.

It is obvious that, although neuroscience has enjoyed tremendous progress, we still do not possess a satisfactory explanation of exactly how brains learn and remember. The hippocampus and related temporal lobe structures are known to be critical for remembering new experiences. But what precisely the hippocampus does, and why an intact hippocampus is required for remembering an event that happened as long as seven weeks previously, remains puzzling. That the hippocampus in some sense "teaches" the cortex, or "prepares" representations for cortical storage seems on the right track. Even if decidely inchoate at this stage, these hunches can be profitably pursued by both modeling and physiological techniques. Does the sleep-dreaming-waking cycle have a role in the hippocampal-cortical connection, as Francis Crick has suggested? How can a neural network learn new things without destroying already stored information?

What are the principles whereby so much detail is discarded in the short term? What determines which aspects of an experience are relevant and should be stored for the long term?

That answers to such questions may very well be forthcoming from convergent research within the next few decades is what makes cognitive neuroscience so tremendously exciting. The excitement also derives from the fact that neuroscience has to do with *us*, and with what makes us what we are. The possibility for practical applications of the knowledge in preventing and treating brain damage is especially poignant, as we daily try to cope with the ravages of diseases such as schizophrenia, stroke, and Huntington's and Alzheimer's disease.

Reference

Churchland PS, Sejnowksi TJ (1991) The net effect: models, methods, and mysteries in computational neuroscience. Cambridge, MIT Press, in press

Temporal Anomalies of Consciousness: Implications of the Uncentered Brain

D. C. Dennett

Summary

As cognitive science, including especially cognitive neuroscience, closes in on the first realistic models of the human mind, philosophical puzzles and problems that have been conveniently postponed or ignored for generations are beginning to haunt the efforts of the scientists, confounding their vision and leading them down hopeless paths of theory. I will illustrate this claim with a brief look at several temporal phenomena which appear anomalous only because of a cognitive illusion: an illusion about the *point of view of the observer.*

Since there is no point in the brain where "it all comes together," several compelling oversimplifications of traditional theorizing must be abandoned.

Wherever there is a conscious mind, there is a *point of view.*[1] This is one of the most fundamental ideas we have about minds, or about consciousness. A conscious mind is an observer, who takes in a limited subset of all the information there is – an observer takes in the information that is available at a particular (roughly) continuous sequence of times and places in the universe. For most practical purposes, we can consider the point of view of a particular conscious subject to be just that: a *point* moving through space-time. We explain the startling time gap between the sound and sight of the distant fireworks by noting the different transmission speeds of sound and light. They arrive *at the observer* (at that point) at different times, even though they left the source at the same time.

What happens, though, when we close in on the observer, and try to locate the observer's *point of view* more precisely, as a point *within* the individual? The simple assumptions that work so well on larger scales begin to break down. There is no single point in the brain where all information funnels in, and this fact has some far-from-obvious indeed quite counterintuitive consequences.

Descartes, one of the first to think seriously about what must happen once we look closely inside the body of the observer, elaborated an idea that is so superficially natural and appealing that it has permeated our thinking about consciousness ever since. Descartes decided that the brain *did* have a center, the

[1] Material in this article is drawn from my forthcoming book. Consciousness Explained. Boston: Little Brown.

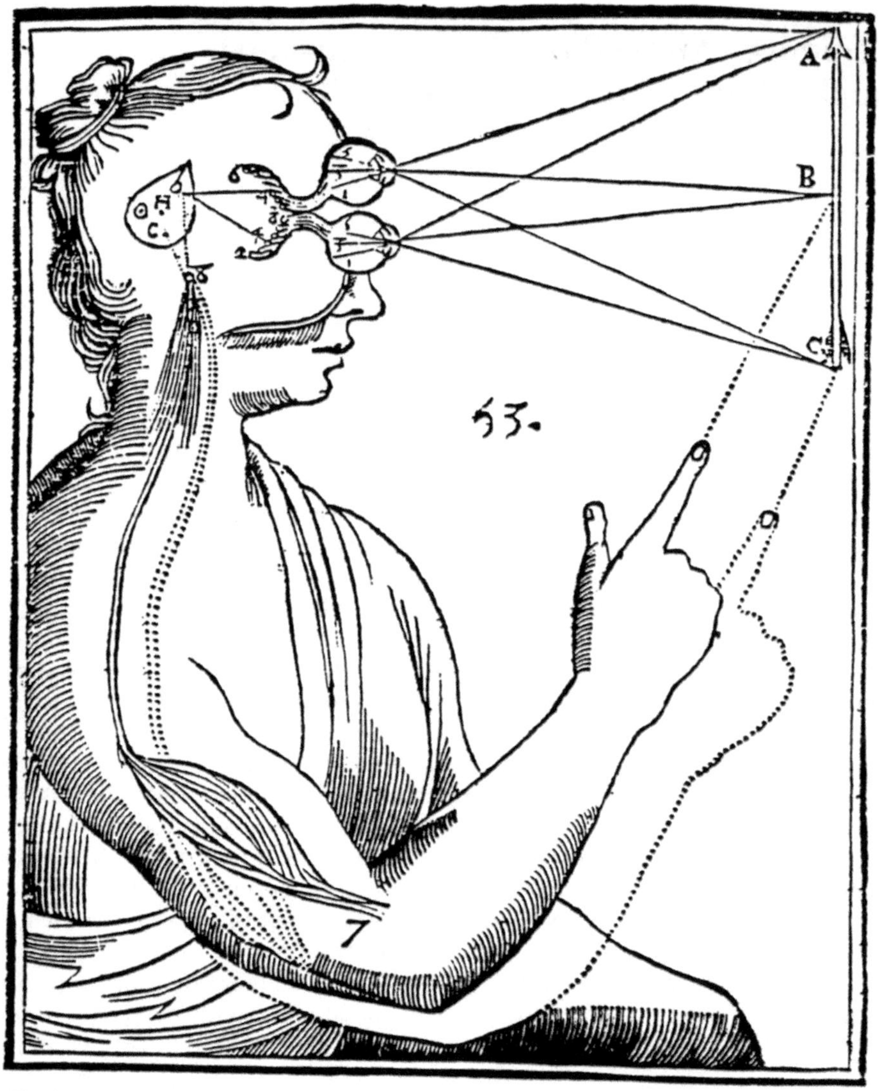

Fig. 1. Descartes' vision of the pineal's role as the turnstile of consciousness

pineal gland, which served as the gateway to the conscious mind (see Fig. 1).

The pineal gland is the only organ in the brain that is in the midline, rather than paired, with left and right versions. Smaller than a pea, it sits in splendid isolation on its stalk, attached to the rest of the nervous system just about in the middle of the back of the brain. Since its function was quite inscrutable (and it still is), Descartes proposed a role for it: in order for a person to be conscious of something, traffic from the senses had to arrive at this station, where it thereupon caused a special, indeed magical transaction to occur between the person's material brain and immaterial mind. Descartes' vision of the pineal's role as the turnstile of consciousness (we might call it the Cartesian bottleneck) is hopelessly wrong. But while materialism of one sort or another is now the received opinion among scientists and philosophers alike, with a few eminent exceptions, even the most sophisticated materialists today often forget that once Descartes' ghostly *res cogitans* is discarded, there is no longer a role for a centralized gateway, or indeed for any *functional* center to the brain. The pineal gland is not only not the FAX machine to the Soul; it is also not the Oval Office of the brain, and neither are any of the other portions of the brain. The brain is Headquarters, the place where the ultimate observer is, but there is no reason to believe that the brain itself has any deeper headquarters, any inner sanctum arrival at which is the necessary of sufficient condition for conscious experience. In short, there is no observer inside the brain.

Light travels much faster than sound, as the fireworks example reminds up, but we now know that it takes longer for the brain to process visual stimuli than to process auditory stimuli. As Ernst Pöppel (1985) has pointed out, thanks to these counterbalancing differences, the "horizon of simultaneity" is about 10 m: light and sound that leave the same point about 10 m from the observer's sense organs produce neural responses that are "centrally available" at the same time. Can we make this figure more precise? There is a problem. The problem is not just measuring the distances from the external event to the sense organs, or the transmission speeds in the various media, or allowing for individual differences. The more fundamental problem is deciding what to count as the "finish line" in the brain.

Pöppel obtained his result by comparing behavioral measures: mean reaction times (button-pushing) to auditory and visual stimuli. The difference ranges between 30 and 40 ms, the time it takes sound to travel approximately 10 m (the time it takes light to travel 10 m is insignificantly different from zero). Pöppel used a peripheral finish line – external behavior – but our natural intuition is that the *experience* of the light and sound happens *between* the time the vibrations hit our sense organs and the time we manage to push the button signalling that experience. And it happens somewhere *centrally*, somewhere in the brain on the excited paths between the sense organ and the finger. It seems that if we could say *exactly* where, we could say exactly when the experience happened. And vice versa: if we could say exactly when it happened, we could say where in the brain conscious experience was located.

We can call the idea of such a centered locus in the brain *"Cartesian materialism"*, since it is the view arrived at when Descartes' dualism is discarded but not the imagery of a central (but material) theater where "it all comes together." Perhaps no one explicitly endorses Cartesian materialism, and no doubt most theorists would insist that they have explicitly rejected such an obviously bad idea. But the persuasive imagery of the Cartesian Theater, in its materialistic form, keeps coming back to haunt them.

This picture of how conscious experience must sit in the brain is a natural extrapolation of the familiar and undeniable fact that for *macroscopic time intervals*, we can indeed order events into the categories "not yet observed" and "already observed" by locating the observer and plotting the motions of the vehicles of information relative to that point. But when we try to extend this method to explain phenomena involving very short time intervals, we encounter a *logical* difficulty. If the "point" of view of the observer is spread over a rather large volume in the observer's brain, the observer's own subjective sense of sequence and simultaneity must be determined by something other than "order of arrival", since order of arrival is incompletely defined until we specify the relevant destination. If A beats B to one finish line but B beats A to another, which result fixes subjective sequence in consciousness? (cf. Minsky 1985). Which point or points of "central availability" would "count" as a determiner of *experienced* order, and why?

Doesn't it follow as a matter of geometric necessity that our conscious minds are located at the *termination* of all the *in*bound processes, just "before" the *initiation* of all the *out*bound processes that implement our actions? Advancing from one periphery along the input channels, we ascend through the optic nerve, and up through various areas of the visual cortex, and then ...? Advancing from the other periphery by swimming upstream from the muscles and the motor neurons that control them, and then up into the supplementary motor area in the cortex and then ...? These two journeys advance towards each other up two slopes, the afferent and the efferent. However difficult it might be to determine in practice the precise location of the Continental Divide in the brain, must there not be, by sheer geometric extrapolation, a highest point, a turning point, a point such that all tamperings on one side of it are *pre-experiental*, and all tamperings on the other are *post-experiential*?

In Descartes's picture, this is obvious to visual inspection, since everything funnels to and from the pineal station. It might seem, then, that if we were to take a more current model of the brain, we should be able to color-code our explorations, using, say, red for afferent or inbound and green for efferent or outbound. Wherever our colors suddenly changed would be a functional midpoint on the great Mental Divide (Fig. 2).

This curiously compelling argument may well ring a bell. It is the twin of an equally spurious argument that has recently been all too influential: Arthur Laffer's notorious Curve, the intellectual foundation (if I may speak loosely) of Reaganomics (Fig. 3).

If the government taxes at 0% it gets no revenue, and if it taxes at 100% no one will work for wages, so it gets no revenue. At 2% it will get roughly twice the

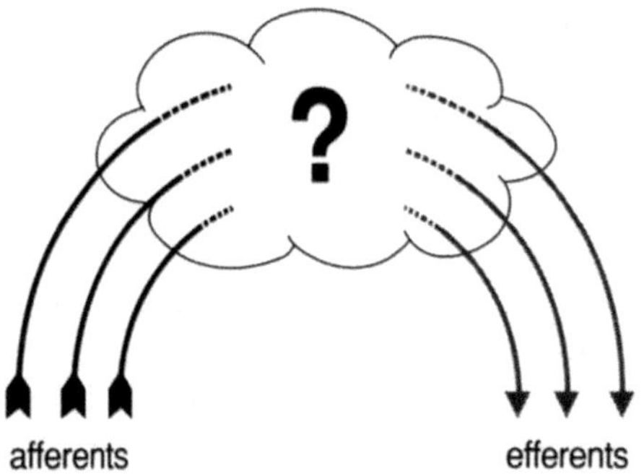

Fig. 2. Afferent/efferent "summit in clouds"

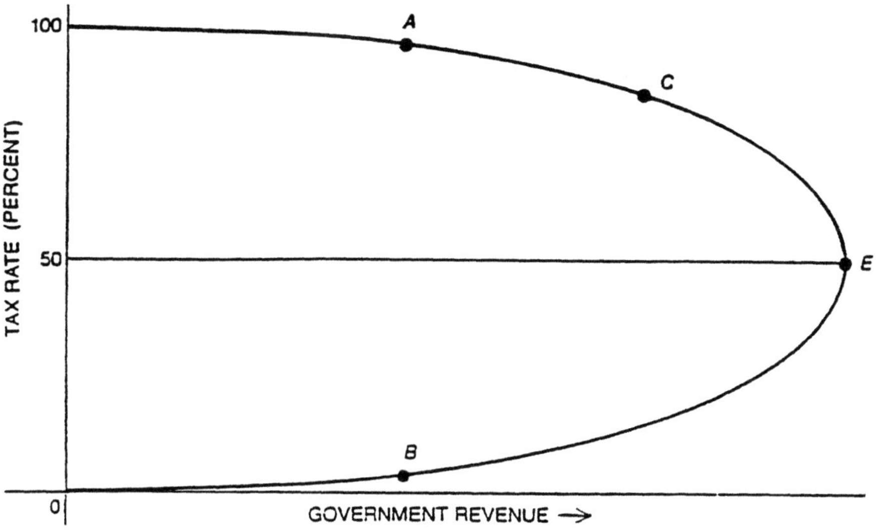

Fig. 3. Laffer curve

revenue as at 1 %, and so forth, but as the rate rises, diminishing returns will set in; the taxes will become onerous. Looking at the other end of the scale, 99 % taxation is scarcely less confiscatory than 100 %, so scarcely any revenue will accrue; at 90 % the government will do better, and better still at the more inviting rate of 80 %. The particular slopes of the curve as shown may be off, but mustn't there be, as a matter of geometric necessity, a place where the curve turns, a rate of taxation that maximizes revenue? Laffer's idea was that, since the current tax rate was on the upper slope, lowering taxes would actually increase revenues. It was a tempting idea. It seemed that it just had to be right. But as Gardner (1981) pointed out, just because the extreme ends of the curve are clear, there is no reason why the unknown part of the curve in the middle regions has to take a smooth course (Fig. 4).

In a satiric mood, he proposed the alternative "neo-Laffer Curve", which has more than one "maximum", and the accessibility of any one of them depends on complexities of history and circumstance that no change of a single variable can possibly determine. We should draw the same moral about what lies in the fog inboard of the afferent and efferent peripheries: the clarity of the peripheries gives us no guarantee that the same distinctions will continue to apply all the way in.

What the "technosnarl" Gardner envisages for the economy is simplicity itself, in fact, compared to the jumble of activities occurring in the more central regions of the brain. So we must stop thinking of the brain as if it had such a central point. This is not just an innocuous shortcut; it is a bad habit. In order to break this bad habit of thought, we need to explore some instances of the bad habit in action, but we also need a good image with which to replace it.

A first version of the good replacement is the Multiple Drafts model of consciousness, but I expect it will seem quite alien and unlikely at first. That's

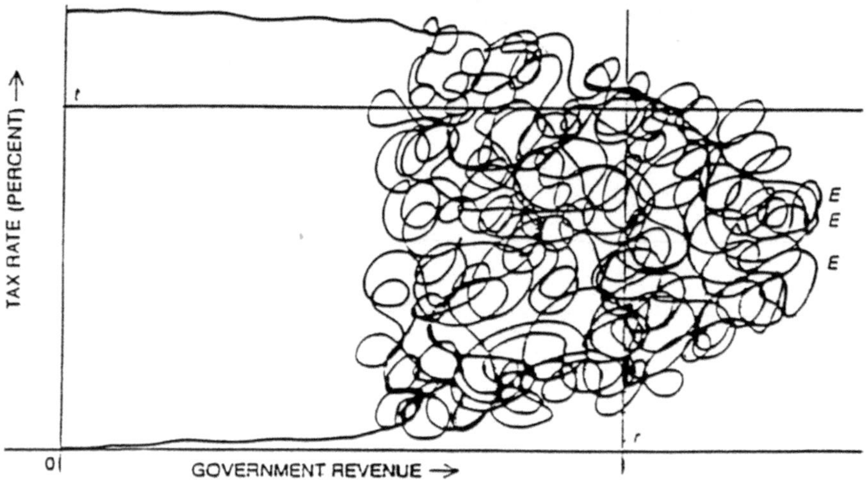

Fig. 4. "NEO-Laffer" curve

how entrenched the Cartesian Theater idea is. According to the Multiple Drafts model, all varieties of perception, and indeed all varieties of thought and action, are accomplished in the brain by multi-track processes of interpretation and elaboration that occur over large fractions of a second, during which time various additions, incorporations, emendations, and overwritings of content can occur, in various orders. Feature-detections or discriminations only have to be made once. That is, once a particular "observation" of some feature has been made by a specialized, localized portion of the brain, the information content thus fixed does not have to be sent somewhere else to be *re*discriminated by some "master" discriminator. In other words, it does not lead to a re-*presentation* of the already discriminated feature for the benefit of the audience in the Cartesian Theater.

These spatially and temporally distributed content-fixations are themselves precisely locatable in both space and time, but their onsets do *not* mark the onset of consciousness of their content. It is always an open question whether any particular content thus discrimated will eventually appear as an element in conscious experience. These distributed content-discriminations yield, over the course of time, something *rather like* a narrative stream or sequence, which can be thought of as subject to continual editing by many processes distributed around in the brain, and continuing indefinitely into the future. This stream of contents is only rather like a narrative because of its multiplicity; at any point in time there are multiple "drafts" of narrative fragments at various stages of "editing" in various places in the brain. Probing this stream at different intervals produces different effects and precipitates different narratives from the subject. If one delays the probe too long (overnight, say) the result is apt to be no narrative left at all, or else a narrative that has been digested or "rationally reconstructed" until it has no integrity. If one probes "too early," one may gather data on how early a particular discrimination is achieved in the stream,

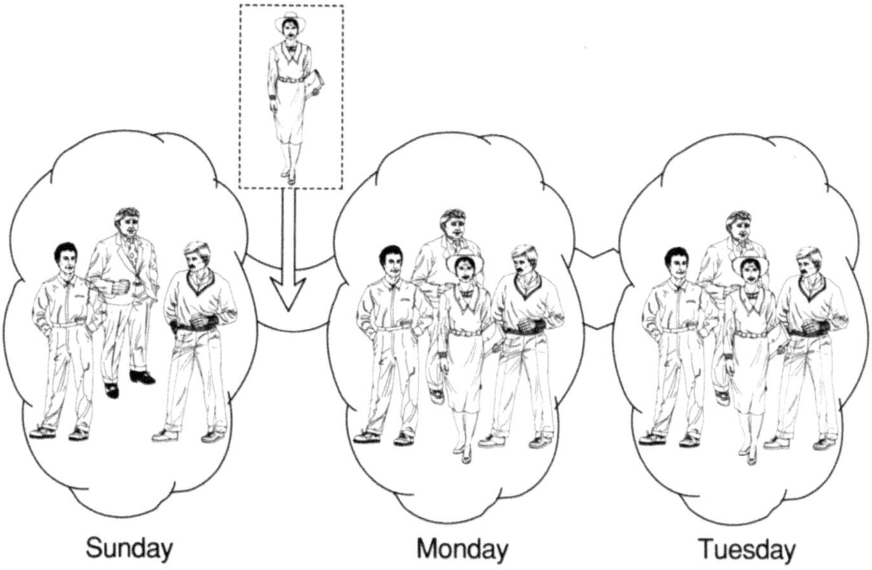

Fig. 5. Lady in the hat at the party

but at the cost of disrupting the normal progression of the stream. Most importantly, the Multiple Drafts model avoids the tempting mistake of supposing that there must be a single narrative (the "final" or "published" draft) that is canonical, the *actual* stream of consciousness of the subject, whether or not the experimenter (or even the subject) can gain access to it. [For a similar model, see William Calvin's 1990 model of consciousness as "scenario-spinning".]

Suppose I tamper with your brain, inserting in your memory a bogus lady in a hat where none was (e.g., at the party on Sunday). If on Monday, when you recall the party, you remember her, and can find no internal resources for so much as doubting the veracity of your memory, we would say that you never did experience her; that is, not at the party on Sunday (Fig. 5).

Of course your subsequent experience of (bogus) recollection can be as vivid as may be, and on Tuesday we can certainly agree that you have had vivid conscious experiences of there being a lady in a hat at the party, but the *first* such experience, we would insist, was on Monday, not Sunday (though it doesn't seem this way to you).

We lack the power to insert bogus memories by neurosurgery, but sometimes our memories play tricks on us, so that what we cannot yet achieve surgically happens in the brain on its own. Sometimes we seem to remember, even vividly, experiences that never occurred. We can call such post-experiential contaminations or revisions of memory *"Orwellian"*, after George Orwell's chilling vision on the novel *1984* of the Ministry of Truth, which busily rewrote history and thus denied access to the (real) past to all who followed.

Orwellian revision is one way to fool posterity. Another is to stage show trials, carefully scripted presentations of false testimony and bogus confessions, complete with simulated evidence. We can call this ploy *"Stalinesque"*. Notice that if we are usually sure which mode of falsification has been attempted on us, the Orwellian or the Stalinesque, this is just a happy accident. In any *successful* disinformation campaign, were we to wonder whether the accounts in the newspapers were Orwellian accounts of trials that never happened at all, or true accounts of phony show trials that actually did happen, we might be unable to tell the difference. If *all* the traces – newspapers, videotapes, personal memoirs, inscriptions on gravestones, living witnesses, etc. – have been either obliterated or revised, we will have no way of knowing whether a fabrication happened *first*, culminating in a staged trial whose accurate history we have before us, or rather *after* a summary execution, history-fabrication covered up the deed, and no trial of any sort actually took place.

The distinction between reality and (subsequent) appearance, and the distinction between Orwellian and Stalinesque methods of producing misleading archives, work unproblematically in the everyday world at macroscopic times scales. One might well think they apply unproblematically *all the way in*, but this is the illusion, and we can catch it in the act in a thought experiment that differs from the first one in nothing but time scale (Fig. 6).

Suppose you are standing on the corner and a long-haired lady dashes by. About one second *after* this, a subterranean memory of some earlier long-haired lady – wearing eyeglasses – contaminates the memory of what you have just seen. When asked a minute later for details of the lady you just saw, you report, sincerely but erroneously, her eyeglasses. Just as in the case of the lady in the hat, we are inclined to say that your original *visual* experience, as opposed to the memory of it seconds later, was *not* of a woman with glasses. But due to the subsequent memory contaminations, it seems to you exactly as if at the first moment you saw her, she surprised you with her eyeglasses. An Orwellian

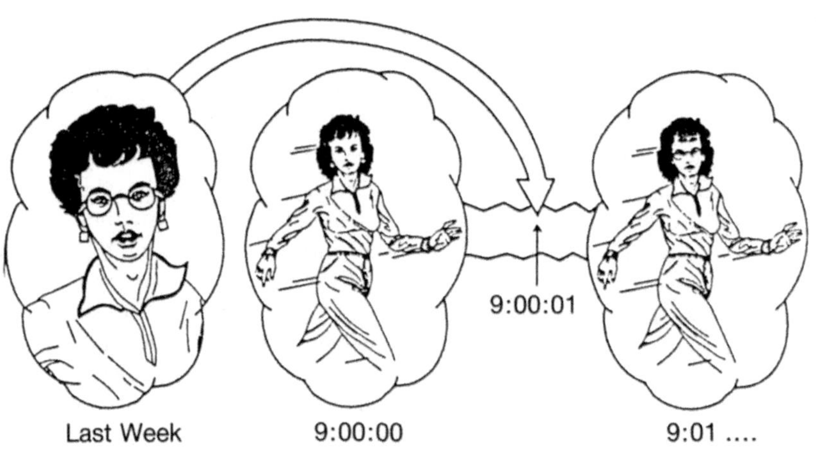

Fig. 6. Lady with glasses/Orwellian version (by Ilil Arbel in Gardner 1981, with permission)

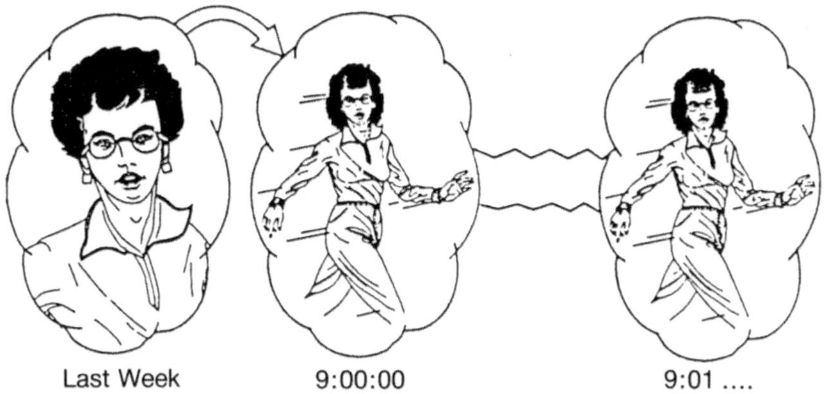

Fig. 7. Lady with glasses/Stalinesque version (by Ilil Arbel in Gardner 1987, with permission)

revision has happened: there was a fleeting instant, before the memory contamination took place, when it *didn't* seem to you that she had glasses on. For that brief moment, the *reality* of your conscious experience was a long-haired lady without eyeglasses. But this historical fact has become inert; it has left no trace, thanks to the contamination of memory that came one second after you glimpsed her.

This understanding of what happened is jeopardized, however, by an alternative account. Your subterranean earlier memories of that long-haired lady with the eyeglasses could just as easily have contaminated your experience *on the upward path*, in the processing of information that occurs "prior to consciousness", so that you actually *hallucinated* the eyeglasses from the very beginning of your experience (Fig. 7).

In that case, your obsessive memory of the lady with the glasses would be playing a Stalinesque trick on you, creating a show trial in experience, which you then accurately recall at later times, thanks to the record in your memory. To naive intuition these two cases are as different as can be: told the first way you suffer no hallucination at the time the lady dashes by, but suffer subsequent memory hallucinations. You have false memories of your actual ("real") experience. Told the second way you hallucinate when she runs by, and then accurately remember that hallucination (which "really did happen in consciousness") thereafter. Surely these are distinct possibilities no matter how finely we divide up time?

No. Here the distinction between perceptual revisions and memory revisions that works so crisply at other scales is not guaranteed application. We have moved into the foggy area (the control window) in which the subject's point of view is spatially and temporally smeared, and the question *"Orwellian or Stalinesque?"* need have no answer.

There is a temporal window that began when the long-haired lady dashed by, exciting your retinas, and ended when you expressed, to yourself or someone else, your eventual conviction that she had eyeglasses on. At some time during this interval, the content *eyeglasses* was spuriously added to the content you expressed. We may assume (and might eventually confirm in detail) that there was a brief time when the content *long-haired lady* had already been discriminated in the brain but before the eyeglasses content had been erroneously "bound" to it. Indeed, it would be plausible to suppose that this discrimination of a long-haired lady was what triggered the memory of the earlier lady with the eyeglasses. What we would not know, however, is whether this spurious binding was before or after the fact the presumed fact of "actual conscious expcrience." Were you first conscious of a lady without eyeglasses and then conscious of a lady with eyeglasses – a subsequent consciousness which wiped out the memory of the earlier experience – or did the very first instant of conscious experience already spuriously contain the glasses? If Cartesian materialism were true, this question would have an answer, even if we, and you, could not determine it retrospectively by any test. But when we abandon Cartesian materialism (as I think we should, and most theorists would agree), the distinction between pre-experiential and post-experiential content revisions cannot always be maintained. I will illustrate this with two simple cases.

Color phi. If two or more small spots separated by as much as 4 degrees of visual angle are briefly lit in rapid succession, a single spot will seem to move. The philosopher Nelson Goodman asked Paul Kolers whether the phi phenomenon persisted if the two illuminated spots were different in color, and if so, what happened to the color of "the" spots as "it" moved? The answer, when Kolers and von Grünau (1976) performed the experiments, was striking: the spot seems to begin moving and then change color abruptly *in the middle of its illusory passage* toward the second location. Goodman (1978) wonders: "how are we able ... to fill in the spot at the intervening place-times along a path running from the first to the second flash *before that second flash occurs?*". (The same question can of course be raised about any phi, but the color-switch in mid-passage vividly brings out the problem.) Unless there is precognition in the brain, the illusory content cannot be created until *after* some identification of the second spot occurs in the brain. But if this identification of the second spot is already "in conscious experience" would it not be too late to interpose the illusory color-switching-while-moving scene between the conscious experience of spot 1 and the conscious experience of spot 2? How does the brain accomplish this sleight-of-hand?

Consider, first, a Stalinesque mechanism In the brain's editing room, located before consciousness, there is a delay, a loop of slack like the "tape delay" used in broadcasts of "live" programs which gives the censors in the control room a few seconds to bleep out obscenities before broadcasting the signal. *In the editing room*, first frame A of the red spot, arrives, then, when frame B of the green spot arrives, some interstitial frames (C and D) can be created and then spliced into the film (in the order A, C, D, B) on its way to projection in the theater of consciousness. By the time the "finished product" arrives at consciousness, it already has its illusory insertion (Fig. 8).

Alternatively, there is an Orwellian mechanism. Shortly after the consciousness of the first spot *and* the second spot (with no illusion of apparent motion at

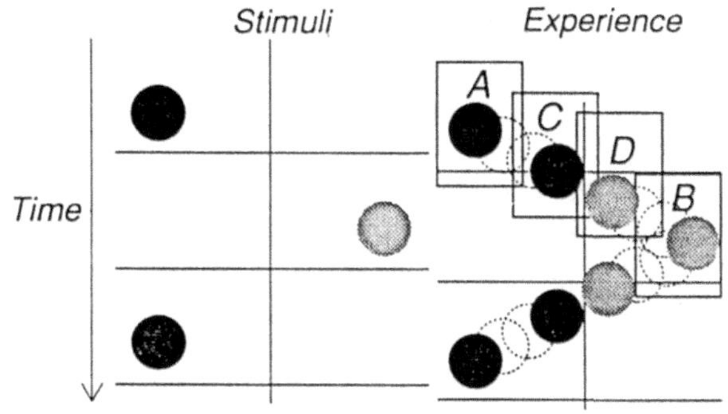

Fig. 8. The phi phenomenon

all), a revisionist historian of sorts, in the brain's memory-library receiving station, notices that the unvarnished history of this incident doesn't make enough sense, so he "interprets" the brute events, red-followed-by-green, by making up a narrative about the intervening passage, complete with midcourse color change. He then installs this history, incorporating his glosses, frames C and D (in Fig. 8), in the memory library for all future reference. Since he works fast, within a fraction of a second (the amount of time it takes to frame (but not utter) a verbal report of what you have experienced), the record you rely on, stored in the library of memory, is already contaminated. You *say* and *believe* that you saw the illusory motion and color change, but that is really a memory hallucination, not an accurate recollection of your original consciousness.

Compare the question here with the question we asked about the lady and the eyeglasses. This time we ask: was the subject conscious of the midcourse color-switch before being conscious of the green spot, or did the color-switch get inserted retrospectively, after consciousness of the green spot had already occurred?

How could we see which of these hypotheses is correct? It might seem that we could rule out the Stalinesque hypothesis quite simply, because of the delay in consciousness it postulates. In Kolers and von Grünau's experiment (1976), there was a 200 ms difference in onset between the red and green spot, and since, *ex hypothesi*, the *whole experience* cannot be composed by the editing room until after the content *green spot* has reached the editing room, consciousness of the initial red spot will have to be delayed by at least that much. (If the editing room sent the content *red spot* up to the theater of consciousness immediately, before receiving frame B and then fabricating frames C and D, the subject would presumably experience a gap in the film, a noticeable delay of around 200 ms between A and C).

Suppose we ask subjects to press a button "as soon as you experience a red spot," and we find little or no difference in response time to a red spot alone versus a red spot followed 200 ms later by a green spot (in which case the subjects report solor-switching apparent motion). This could be because there is *always* a delay of at least 200 ms in consciousness, but aside from the biological implausibility of such a squandering of time, there is the evidence from many quarters that responses under conscious control, while slower than such responses as reflex blinks, occur with close to the minimum latencies that are physically possible. After subtracting the demonstrable travel times for incoming and outgoing pulse trains, and the response preparation time, there is little time left over in "central processing" in which to hide a 200 ms delay. So the responses had to have been initiated before the discrimination of the second stimulus, the green spot. This would seem overwhelmingly to favor the Orwellian, post-experiential mechanism: as soon as the subject *becomes conscious* of the red spot, he initiates a button-press. *While that button press is forming*, he becomes conscious of the green spot. *Then* both these experiences are wiped from memory, replaced in memory by the revisionist record of the red spot moving over and then turning green halfway across. He readily and sincerely (but falsely) reports having seen the red spot moving towards the green spot before changing color.

If the subject insists that he really was conscious from the very beginning of the red spot moving and changing color, the Orwelian theorist will firmly explain to him that he is wrong. His memory is playing tricks on him. The fact that he pressed the button when he did is conclusive evidence that he was conscious of the (stationary) red spot before the green spot had even occurred. After all, his instructions were to press the button *when he was conscious of a* red spot. He must have been conscious of the red spot about 200 ms before he could have been conscious of it moving and turning green. If that is not how it seems to him, he is simply mistaken.

The defender of the Stalinesque (pre-experiential) alternative is not defeated by this, however. Actually, he insits the subject responded to the red spot *before* he was conscious of it! The directions to the subject (to respond to a red spot) and somehow trickled down from consciousness into the editing room, which *unconsciously* initiated the button-push before sending the edited version (frames ACDB) up to consciousness for "viewing". The subject's memory has played no tricks on him. He is reporting exactly what he was conscious of, except for his insistence that he consciouly pushed the button after seeing the red spot; his "premature" button-push was unconsciously (or preconsciously) triggered.

Where the Stalinesque theory postulates a button-pushing reaction to an *un*conscious detection of a red spot, the Orwellian theory postulates a *conscious* experience of a red spot that is immediately obliterated from memory by its sequel. So here is the rub: we have two different models of what happens in the phi phenomenon. One posits a Stalinesque "filling in" on the upward, pre-experiential path, and the other posits an Orwellian "memory revision" on the downward, post-experiental path, and *both* of them are consistent with *whatever* the subjects says or thinks or remembers. Note that the inability to distinguish between these two possibilities does not just apply to the *outside* observers who might be supposed to lack some private data to which the subject had "privileged access." You, as a subject in a phi phenomenon experiment, *could not* discover anything in the experience from your own first-person perspective that would favor one theory over the other; the experience would "feel the same" on either account.

Today we have grown quite comfortable with the distinction between the spatial location in the brain of the vehicle of experience and the location "in experiential space" of the item experienced. In short we distinguish representing from represented, vehicle from content. We have grown sophisticated enough to recognize that the products of visual perception are not, literally, pictures in the head even though *what they represent* is what pictures represent well: the layout in space of various visible properties. What we tend to miss is that we should make the same distinction for time: *when* in the brain an experience happens must be distinguished from when it seems to happen. Indeed, one way of looking at the message of this chapter is as a straightforward extension of the common wisdom about experience of space to experience of time. The representation of space in the brain does not always use space-in-the-brain to represent space, and the representation of time in the brain does not always use time-in-the-brain. In fact, it *cannot* always use time-in-the-brain because of the lack of a single finish line, or Cartesian theater, where "order of arrival" fixes subjective order in consciousness.

References

Calvin W (1990) The cerebral symphony: seashore reflections on the structure of consciousness. Bantam, New York

Gardner M (1981) The Laffer curve and other laughs in current economics. Sci Am 245: 18–31

Goodman N (1978) Ways of worldmaking. Harvester, Hassocks

Kolers P, von Grünau M (1976) Shape and color in apparent motion. Vision Res 16: 329–335

Minsky M (1985) The society of mind. Simon and Schuster

Pöppel E (1985) Grenzen des Bewußtseins. Deutsche Verlags-Anstalt Stuttgart. Translated as Mindworks: time and conscious experience (1988) Harcourt Brace Jovanovich, New York

Intertheoretic Reduction: A Neuroscientist's Field Guide

P. M. Churchland and *P. S. Churchland*

Summary

Might psychology someday be reduced to (that is, be exhaustively explained by) computational neurobiology? Many still say no. We approach this question through a brief survey of some prominent intertheoretic reductions drawn from the history of science. A general characterization of reduction is constructed from these, and some important philosophical and methodological lessons are drawn. The five most popular objections to the possibility of a neurobiological reduction of psychology are then addressed, and defeated.

Introduction

'Reductionism' is a term of contention in academic circles. For some, it connotes a right-headed approach to any genuinely scientific field, an approach that seeks intertheoretic unity and real systematicity in the phenomena. It is an approach to be vigorously pursued and defended.

For others, it connotes a wrong-headed approach that is narrow-minded and blind to the richness of the phenomena. It is a bullish instance of ‚nothing-butery', insensitive to emergent complexity and higher-level organization. It is an approach to be resisted.

This latter reaction is most often found within the various social sciences, such as anthropology, sociology, and psychology. The former attitude is most often found within the physical sciences, such as physics, chemistry, and molecular biology. Predictably then, the issue of reductionism is especially turbulent at the point where these two intellectual rivers meet: in the discipline of modern neuroscience.

The question at issue is whether it is reasonable to expect, and to work toward, a reduction of all psychological phenomena to neurobiological and neurocomputational phenomena. A large and still respectable contingent within the academic community remains inclined to say no. Their resistance is principled. Some point to the existence of what philosophers call *qualia* – the various subjective qualitative characters displayed in our sensations: think of pain, the

This text was published in Seminars in the Neurosciences 1990, vol. 2, pp 249–256.

smell of a rose, the sensation of redness, and so forth. These qualia, it is held, are beyound the possibility of any materialist explanation or reduction (Jackson 1982, Nagel 1974). Others point to the semantic content or *intentionality* of our thoughts, and make a similar claim about its irreducibility (Popper and Eccles 1978; Searle 1990). Others claim that the most important aspects of human behavior are explicable only in terms of high-level *emergent properties* and their correlative regularities, properties that irreducibly encompass the social level, properties such as loyalty to a moral ideal, perception of a political fact, or the recognition of a personal betrayal (Taylor 1970, 1987). Yet others see a conflict with the important and deeply-entrenched idea of *human freedom* (Popper and Eccles 1978). Finally, some materialists raise what is called the problem of *multiple instantiation.* They point to the presumed fact that conscious intelligence could be sustained by physical systems other than the biochemistry peculiar to humans – by a system of transistors, for example – just as a nation's financial economy can be substained by tokens other than silver coins and paper bills. But no one thinks that macroeconomics can be reduced to the chemistry of metals and paper. So why think that psychology should be reducible to the neurobiology of terrestrial humans (Fodor 1975)?

Our aim here is three-fold. First, we will try to provide a useful overview of the general nature of intertheoretic reduction, as it appears in the many examples to be found in the history of science. Expanding our horizons here is important, since little is to be learned from simply staring long and hard at the problematic case at issue, namely, the potential reduction of psychological phenomena to neural phenomena. Instead, we need to look at cases where the dust has already settled and where the issues are already clear. Second, we will identify the very real virtues that such cases display, and the correlative vices to be avoided. And finally, we will attempt to apply these historical lessons to the case here at issue – cognitive neuroscience – and we will try to meet the salient objections listed above.

Intertheoretic Reduction: Some Prototypical Cases

Since nothing instructs like examples, let us briefly examine some. One of the earliest cases of intertheoretic reduction on a grand scale was the reduction of Kepler's three laws of astronomical motion by the newly minted mechanics of Isaac Newton. Kepler's theory was specific to the motions of the solar planets,

> Kepler's three planetary laws are: all planets move on ellipses with the Sun at one focus;
> a given planet always sweeps out equal areas in equal times;
> the square of planet's period is proportional to the cube of its mean orbital radius.
> Newton's three laws of motion are:
> inertial motion is constant and rectilinear;
> Acceleration = force/mass;
> for any change in momentum something suffers an equal and opposite change in momentum. To these laws we must add his gravitation law: $F = Gm_1m_2/R^2$.

but Newton's theory at least purported to be the correct account of bodily motions in general. It was therefore a great triumph when Newton showed that one could deduce all three of Kepler's laws from his own theory, given only the background assumption that the mass of any planet is tiny compared to the great mass of the Sun.

Kepler's account thus turned out to be just a special case or a special application of Newton's more encompassing account. And astronomical motions turned out to be just a special instance of the inertial and force-governed motions of massive bodies in general. The divine or supernatural character of the heavens was thereby lost forever. The sublunary and the super-lunary realms were thereby united as a single domain in which the same kinds of objects were governed by one and the same set of laws.

Newton's mechanics also provides a second great example of intertheoretic reduction, one that did not emerge until the nineteenth century. If his mechanics successfully comprehends motion at both the astronomical and the human-sized scales, then what, it was asked, about motions at the *microscopic* scale? Might these be accounted for in the same way?

The attempts to construct such an account produced another unification, one with an unexpected bonus concerning the theory of heat. If we assume that any confined body of gas consists in a swarm of submicroscopic corpuscles bouncing around inside the container according to Newton's three laws, then we can deduce a law describing the pressure they will collectively exert on the container's walls by repeatedly bouncing off them. This ‚kinetic' law has the form

$$PV = 2n/3 \cdot mv^2/2$$

This law had the same form as the then already familiar 'ideal gas law',

$$PV = \mu R \cdot T$$

(Here P is pressure and V is volume). Although they are notationally different, the expressions '$2n/3$' and 'μR' both denote the amount of gas present in the container ('n' denotes the number of molecules in the container; 'μ' denotes the fraction of a mole). The only remaining difference, then, is that the former law has an expression for the *kinetic energy of an average corpuscle* ($mv^2/2$) in the place where the latter has an expression for *temperature* (T). Might the phenomenon we call 'temperature' thus *be* mean kinetic energy at the molecular level? This striking convergence of principle, and many others like it, invited Bernoulli, Joule, Kelvin, and Boltzmann to say yes. As matters were further pursued, mean molecular kinetic energy turned out to have *all* of the causal properties that the classical theory had been ascribing to temperature. In short, temperature turned out to *be* mean molecular kinetic energy. Newtonian mechanics had another reductive triumph in hand. Motion at all three scales was subsumed under the same theory, and a familiar phenomenal property, *temperature*, was reconceived in a new and unexpected way.

It is worth emphasizing that this reduction involved identifying a familiar *phenomenal* property of common objects with a highly unfamiliar microphysical property. (By 'phenomenal', we mean a property one can reliably discriminate in experience, but where one is unable to articulate, by reference to yet simpler discriminable elements, just how one discriminates that property.) Evidently, reduction is not limited to conceptual frameworks hidden away in the theoretical stratosphere. Sometimes the conceptual framework that gets subsumed by a deeper vision turns out to be a familiar piece of our common-sense framework, a piece whose concepts are regularly applied in casual observation on the basis of our native sensory systems. Other examples are close at hand: before Newton, *sound* had already been identified with compression waves in the atmosphere, and *pitch* with wavelength, as part of the larger reduction of common-sense sound and musical theory to mechanical acoustics. A century and a half after Newton, *light* and its various *colors* were identified with electromagnetics waves and their various wavelenghts, within the larger reduction of geometrical optics by electromagnetic theory, as outlined by Maxwell in 1864. *Radiant heat,* another common-sense observable, was similarly reconceived as long-wavelength electromagnetic waves in a later articulation of the same theory. Evidently, the fact that a property or state is at the prime focus of one of our native discriminatory faculties does not mean that it is exempt from possible reconception within the conceptual framework of some deeper explanatory theory.

This fact will loom larger later in the paper. For now, let us explore some further examples of intertheoretic reduction. The twentieth century reduction of classical (valence) chemistry by atomic and subatomic (quantum) physics is another impressive case of conceptual unification. Here the structure of an atom's successive electron shells, and the character of stable regimes of electron-sharing between atoms, allowed us to reconstruct, in a systematic and thus illuminating way, the electronic structure of the many atomic elements, the classical laws of valence-bonding, and the gross structure of the periodic table. As often happens in intertheoretic reductions, the newer theory also allowed us to explain much that the old theory had been unable to explain, such as the specific heat capacities of various substances and the interactions of chemical compounds with light.

This reduction of chemistry to physics is notable for the further reason that it is not yet complete, and probably never will be. For one thing, given the combinatorial possibilities here, the variety of chemical compounds is effectively endless, as are their idiosyncratic chemical, mechanical, optical, and thermal properties. And for another, the calculation of these diverse properties from basic quantum principles is computationally daunting, even when we restrict ourselves to merely approximate results, which for sheerly mathematical reasons we generally must. Accordingly, it is not true that all chemical knowledge has been successfully reconstructed in quantum-mechanical terms. Only the basics have, and then only in approximation. But our experience here firmly suggests that quantum physics has indeed managed to grasp the underlying elements of chemical reality. We thus expect that any particular *part* of chemistry can be approximately reconstructed in quantum mechanical terms, when and if the specific need arises.

The preceding examples make it evident that intertheoretic reduction is at bottom a relation between two distinct *conceptual frameworks* for describing the phenomena, rather than a relation between two distinct domains of phenomena. The whole point of a reduction, after all, is to show that what we thought to be two domains is actually one domain, though it may have been described in two (or more) different vocabularies.

Perhaps the most famous reduction of all is Einstein's twentieth century reduction of Newton's three laws of motion by the quite different mechanics of the Special Theory of Relativity (STR). STR subsumed Newton's laws in the following sense. If we make the (false) assumption that all bodies move with velocities much less than the velocity of light, then STR entails a set of laws for the motion of such bodies, a set that is experimentally indistinguishable from Newton's old set. It is thus no mystery that those old Newtonian laws seemed to be true, given the relatively parochial human experience they were asked to account for.

But while those special-case STR laws may be experimentally indistinguishable from Newton's laws, they are logically and semantically quite different from Newton's laws: they ascribe an importantly different family of features to the world. Specifically, in every situation where Newton ascribed an intrinsic property to a body (e.g. mass, or length, or momentum, and so forth), STR ascribes a *relation, a two*-place property (e.g. *x* has a mass-relative-to-an-inertial-frame-F, and so on), because its portrait of the universe and what it contains (an unitary 4-D spacetime continuum with 4-D world-lines) is profoundly different from Newton's.

Here we have an example where the special-case resources and deductive consequences of the new and more general theory are not identical, but merely *similar*, to the old and more narrow theory it purports to reduce. That is to say, the special-case reconstruction achieved within the new theory parallels the old theory with sufficient systematicity to explain why the old theory worked as well as it did in a certain domain, and to demonstrate that the old theory could be displaced by the new without predictive or explanatory loss within the old theory's domain; and yet the new reconstruction is not perfectly isomorphic to the old theory. The old theory turns out not just to be narrow, but to be false in certain important respects. Space and time are not distinct, as Newton assumed, and there simply are no intrinsic properties such as mass and length that are invariant over all inertial frames.

The trend of this example leads us toward cases where the new and more general theory does not sustain the portrait of reality painted by the old theory at all, even as a limiting special case or even in its roughest outlines. An example would be the outright displacement, without reduction, of the old phlogiston theory of combustion by Lavoisier's oxygen theory of combustion. The older theory held that the combustion of any body involved the *loss* of a spirit-like substance, phlogiston, whose pre-combustion function it was to provide a noble wood- or metl-like character to the baser ash or calx that is left behind after the process of combustion is complete. It was the 'ghost' that gave metal its form. With the acceptance of Lavoisier's contrary claim that a sheerly material substance, oxygen, was being somehow *absorbed* during combustion, phlogiston was simply eliminated from our overall account of the world.

Other examples of theoretical entities that have been eliminated from serious science include caloric fluid, the rotating crystal spheres of Ptolemaic astronomy, the four humors of medieval medicine, the vital spirit of premodern biology, and the luminiferous aether of pre-Einsteinian mechanics. In all of these cases, the newer theory did not have the resources adequate to reconstruct the furniture of the older theory or the laws that supposedly governed their behavior; but the newer theory was so clearly superior to the old as to displace it regardless.

At one end of the spectrum then, we have pairs of theories where the old is smoothly reduced by the new, and the ontology of the old theory (that is, the set of things and properties that is postulates) survives, although redescribed, perhaps, in a new and more penetrating vocabulary. Here we typically find claims of cross-theoretic identity, such as 'Heat is identical with mean molecular kinetic energy' and 'Light is identical with electromagnetic waves'. In the middle of the spectrum, we find pairs of theories where the old ontology is only poorly mirrored within the vision of the new, and it ‚survives' only in a significantly modified form. Finally, at the other end of the spectrum we find pairs where the older theory, and its ontology with it, is eliminated entirely in favor of the more useful ontology and the more successful laws of the new.

Before closing this quick survey, it is instructive to note some cases where the older theory is neither subsumed under nor eliminated by the aspirant and allegedly more general theory. Rather, it successfully resists the takeover attempt, and proves not to be just a special case of the general theory at issue. A clear example is Maxwell's electromagnetic theory (hereafter, EM theory). From 1864 to 1905, it was widely expected that EM theory would surely find a definitive reduction in terms of the mechanical properties of an all-pervading aether, the elastic medium in which EM waves were supposedly propagated. Though never satisfactorily completed, some significant attempts at reconstructing EM phenomena in mechanical terms had already been launched. Unexpectedly, the existence of such an absolute medium of luminous propagation turned out to be flatly inconsistent with the character of space and time as described in Einstein's 1905 Special Theory of Relativity. EM theory thus emerged as a fundamental theory in its own right, and not just as a special case of mechanics. The attempt at subsumption was a failure.

A second example concerns the theory of stellar behavior accumulated by classical astronomy in the late nineteenth century. It was widely believed that the pattern of radiative behavior displayed by a star would be adequately explained in mechanical and/or in chemical terms. It became increasingly plain, however, that the possible sources of chemical and mechanical energy available to any star would sustain their enormous outpourings of thermal and luminous energy for only a few tens of millions of years. This limited time scale was at odds with the emerging geological evidence of a history numbered in the *billions* of years. Geology notwithstanding, Lord Kelvin himself was prepared to bite the bullet and declare the stars to be no more than a few tens of millions of years old. The conflict was finally resolved when the enormous energies in the atomic nucleus were discovered. Stellar astronomy was eventually reduced all right, and very beautifully, but by quantum physics rather than by mere chemistry or mechanics. Another reductive attempt had failed, though it was followed by one that succeeded.

The Lessions for Neuroscience

Having seen these examples and the spectrum of cases they define, what lessons should a neuroscientist draw? One lession is that intertheoretic reduction is a normal and fairly commonplace event in the history of science. Another lesson is that genuine reduction, when you can get it, is clearly a *good* thing. It is a good thing for many reasons, reasons made more powerful by their conjunction.

First, by being displayed as a special case of the (presumably true) new theory, the old theory is thereby *vindicated,* at least in its general outlines, or at least in some suitably restricted domain. Second, the old theory is typically *corrected* in some of its important details, since the reconstructed image is seldom a perfect mirror image of the old theory, and the differences reflect improvements in our knowledge. Third, the reduction provides us with a much *deeper insight* into, and thus a *more effective control* over, the phenomena within the old theory's domain. Fourth, the reduction provides us with a *simpler* overall account of nature, since apparently diverse phenomena are brought under a single explanatory umbrella. And fifth, the new and more general theory immediately *inherits all the evidence* that had accumulated in favor of the older theory it reduces, because it explains all of the same data.

It is of course a bad thing to try to force a well-functioning old theory into a procrustean bed, to try to effect a reduction where the aspirant reducing theory lacks the resources to do reconstructive justice to the target old theory. But whether or not the resources are adequate is seldom clear beforehand, despite people's intuitive convictions. And even if a reduction is impossible, this may reflect the old theory's radical falsity instead of its fundamental accuracy. The new theory may simply eliminate the old, rather than smoothly reduce it. Perhaps folk notions such as 'beliefs' and 'the will', for example, will be eliminated in favor of some quite different story of information storage and behavior initiation.

The fact is, in the neuroscience/psychology case there are conflicting indications. On the one side, we should note that the presumption in favor of an eventual reduction (or elimination) is far stronger than it was in the historical cases just examined. For unlike the earlier cases of light, or heat, or heavenly motions, in general terms we already know how psychological phenomena arise: they arise from the evolutionary and ontogenetic articulation of matter, more specifically, from the articulation of biological organization. We therefore *expect* to understand the former in terms of the latter. The former is produced by the relevant articulation of the latter.

But there are counter indications as well, and this returns us at last to the five objections with which we opened this paper. From the historical perspective outlined above, can we say anything useful about those objections to reduction? Let us take them in sequence.

The first concerns the possibility of explaining the character of our subjective sensory qualia. The negative arguments here all exploit the very same theme, *viz* our inability to imagine how any possible story about the objective nuts and bolts of neurons could ever explain the inarticulable subjective phenomena at issue. Plainly this objection places a great deal of weight on what we can and cannot imagine, as a measure of what is and isn't possible. It places more, clearly, than the test should bear. For who would have imagined, before James Clark Maxwell, that the theory of charged pith balls and wobbling compass needles could prove adequate to explain all the phenomena of light? Who would have thought, before Descartes, Bernoulli and Joule, that the mechanics of billiard balls would prove adequate to explain the *prima facie* very different phenomenon of heat? Who would have found it remotely plausible that the pitch of a sound is a frequency, in advance of a general appreciation that sound itself consists in a train of compression waves in the atmosphere?

We must remember that a successful intertheoretic reduction is typically a complex affair, as it involves the systematic reconstruction of all or most of the old conception within the resources of the new conception. And not only is it complex, often the reconstruction is highly surprising. It is not something that we can reasonably expect anyone's imagination to think up or comprehend on rhetorical demand, as in the question, 'How could A's *possibly* be nothing but B's?'

Besides, this rhetorical question need not stump us if our imagination is informed by recent theories of sensory coding. The idea that taste sensations are coded as a four-dimensional vector of spiking frequencies (corresponding to the four types of receptor on the tongue) yields a representation of the space of humanly possible tastes which unites the familiar tastes according to their various similarities, differences, and other relations such as betweenness (Bartoshuk 1978). Land's retinex theory of color vision (Land 1977) suggests a similar arrangement for our color sensations, with similar virtues. Such a theory also predicts the principal forms of color blindness, as when one's three-dimensional color space is reduced to two dimensions by the loss of one of the three classes of retinal cones.

Here we are already reconstructing some of the features of the target phenomena in terms of the new theory. We need only to carry such a reconstruction through, as in the historical precedents of the objective phenomenal properties noted earlier (heat, light, pitch). Some things may indeed be inarticulably phenomenal in character, because they are the target of one of our basic discriminatory modalities. But that in no way makes them immune to an illuminating intertheoretic reduction. History already teaches us the contrary.

The second objection concerned the meaning, or semantic content, or intentionality of our thoughts and other mental states. The anti-reductionist arguments in this area are very similar to those found in the case of qualia. They appeal to our inability to imagine how meaning could be just a matter of how signals interact or how inert symbols are processed (Searle 1980, 1990; for a rebuttal, see Churchland and Churchland 1990). Searle, strictly speaking, objects only to a purely computational reduction, but that is an important option for neuroscience so we shall include him with the other anti-reductionists. Such appeals, as before, are really arguments from ignorance. They have the form, 'I can't *imagine* how a neurocomputational account of meaningful representations could possibly work; therefore, it can't possibly work'. To counter such appeals in the short-term, we need only point out this failing.

To counter them in the long-term requires more. It requires that we actually produce an account of how the brain represents the external world and the regularities it displays. But that is precisely what current theories of neural network function address. According to them, real-time information about the world is coded in high-dimensional activation vectors, and general information about the world is coded in the background configuration of the network's synaptic weights. Activation vectors are processed by the weight-configurations through which they pass, and learning consists in the adjustment of one's global weight configuration (see Fig. 1)

These accounts already provide the resources to explain a variety of things, such as the recognition of complex objects despite partial or degraded sensory inputs, the swift retrieval of relevant information from a vast content-addressable memory, the appreciation of diffuse and inarticulable similarities, and the administration of complex sensorimotor coordination (Churchland 1989). We are still too ignorant to insist that hypotheses of this sort will prove adequate to explain all of the representational capacities of mind. But neither can we insist that they are doomed to prove inadequate. It is an empirical question, and the jury is still out.

The third objection complains that what constitutes a human consciousness is not just the intrinsic character of the creature itself, but also the rich matrix of relations it bears to the other humans, practices, and institutions of its embedding culture. A reductionistic account of human consciousness and behavior, insofar as it is limited to the microscopic activities in an individual's brain, cannot hope to capture more than a small part of what is explanatorily important.

The proper response to this objection is to embrace it. Human behavior is indeed a function of the factors cited. And the character of any individual human consciousness will be profoundly shaped by the culture in which it develops. What this means is that any adequate neuro-computational account of human consciousness must take into account the manner in which a brain comes to represent, not just the gross features of the physical world, but also the character of the other cognitive creatures with which it interacts, and the details of the social, moral, and political world in which they all live. The brains of social animals, after all, learn to be interactive elements in a community of brains, much to their cognitive advantage. We need to know how they do it.

This is a major challenge, one that neuroscientists have not yet addressed with any seriousness, nor even much acknowledged. This is not surprising. Accounting for a creature's knowledge of the spatial location of a fly is difficult enough. Accounting for its knowledge of a loved one's embarrassment, a politician's character, or a bargaining opponent's hidden agenda, represents a much higher level of difficulty. And yet we already know that artificial neural networks, trained by examples, can come to recognize and respond to the most astonishingly subtle patterns and similarities in nature. If physical patterns, why

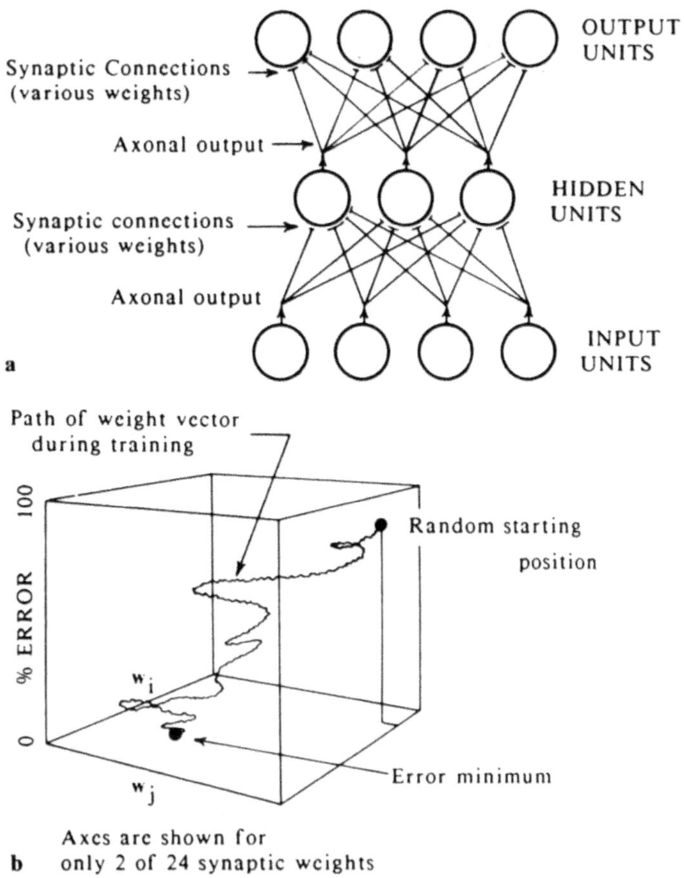

Fig. 1a,b. **a** A simple feedforward artificial 'neural network'. Inputs are coded as a pattern or vector of activation levels across the input units. This pattern is conveyed toward the hidden units, but is transformed as it passes through the bank of intervening synaptic connections of various weights. Each hidden unit then sums the weighted inputs it receives and assumes an activation level appropriate to that sum. Thus results a second pattern or vector of activation levels across the hidden units. The story is repeated for the final layer of output units, which assumes a third pattern of activations. The network is thus a vector-to-vector transformer. Precisely what transformation it embodies is dictated by the specific configuration of its synaptic weights. With sigmoid output functions at the first two layers, networks of this kind can approximate any computable transformation whatever. **b** A schematic portrayal of an abstract ,weight space' for the simple network. Learning in such a network consists in the successive modification of its weight configuration inorder to incrementally reduce its performance error. This process continues until the network finally performs the input/output transformation implicit in the many examples on which it was trained. (From Churchland 1989)

not social patterns? We confront no problem in principle here. Only a major challenge.

It may indeed be unrealistic to expect an exhaustive global account of the neural and behavioral trajectory of a specific person over any period of time. The complexity of the neural systems we are dealing with may forever preclude anything more than useful approximations to the desired ideal account. The case of chemistry and its relation to quantum physics comes to mind. There also, the mathematics of complex dynamical systems imposes limits on how easily and accurately we can reconstruct the chemical facts from the physical principles. This means that our reduction will never be truly complete, but we rightly remain confident that chemical phenomena are nothing but the macro-level reflection of the underlying quantum physical phenomena even so. As with chemical phenomena, so with psychological phenomena.

This brings us to the fourth objection, concerning the threat that a reduction would pose to human freedom. Here we shall be brief. Whether and in what sense there is any human freedom, beyond the relative autonomy that attaches to any complex dynamical system that is partially isolated from the world, is an entirely empirical question. Accordingly, rather than struggle to show that a completed neuroscience will be consistent with this, that, or the other preconceived notion of human freedom, we recommend that we let scientific investigation *teach us* in what ways and to what degrees human creatures are 'free'. No doubt this will entail modifications for some people's current conceptions of human freedom, and the complete elimination of some others. But that is preferable to making our current confusions into a standard that future theories must struggle to be consistent with.

The fifth and final objection claims an irreducibly abstract status for psychology, on grounds that a variety of quite different physical systems could realize equally well the abstract organization that constitutes a cognitive economy. How can we reduce psychological phenomena to neurobiology, if other physical substrates might serve just as well?

The premise of this objection will likely be conceded by all of us. But the conclusion against reduction does not follow. We can see this clearly by examining a case from our own scientific history. Temperature, we claimed earlier, is identical with mean molecular kinetic energy. But strictly speaking, this is true only for a gas, where the molecules are free to move in a ballistic fashion. In a solid, where the particles oscillate back and forth, their energy is constantly switching between a kinetic and a potential mode. In a high-temperature plasma, there are no molecules at all to consider, since everything has been ripped into sub-atomic parts. Here temperature is a complex mix of various energies. And in a vacuum, where there is no mass at all, temperature consists in the wavelength distribution – the 'black-body curve' – of the EM waves passing through it.

What these examples show us is that reductions can be domain specific: in a gas, temperature is one thing; in a solid, temperature is another thing; in a plasma, it is a third; in a vacuum, a fourth; and so on. (They al count as 'temperatures', since they interact, and they all obey the same laws of equilibrium and disequilibrium). None of this moves us to say that classical thermodynamics is an autonomous, irreducible science, forever safe from the ambitions of the underlying microphysical story. On the contrary, it just teaches us that there is more than one way in which energy can be manifested at the microphysical level.

Similarly, visual experience may be one thing in a mammal, and a slightly different thing in an octopus, and a substantially different thing in some possible metal-and-semiconductor android. But they will all count as visual experiences because they share some set of abstract features at a higher level of description. That neurobiology should prove capable of explaining all psychological phenomena in humans is not threatened by the possibility that some *other* theory, say semiconductor electronics, should serve to explain psychological phenomena in *robots*. The two reductions would not conflict. They would complement each other.

We have elsewhere provided more comprehensive accounts of how recent work in neuroscience illuminates issues in psychology and cognitive theory (Churchland 1986; Churchland 1989). We conclude here with two cautionary remarks. First, while we have here been very upbeat about the possibility of reducing psychology to neuroscience, producing such a reduction will surely be a long and difficult business. We have here been concerned only to rebut the counsel of impossibility, and to locate the reductive aspirations of neuroscience in a proper historical context.

Second, it should be not assumed that the science of psychology will somehow disappear in the process, nor that its role will be limited to that of a passive target of neural explanation. On the contrary, chemistry has not disappeared through the quantum-mechanical explication of its basics; nor has the science of biology disappeared, despite the chemical explication of its basics. Moreover, each of these higher-level sciences has helped to shape profoundly the development and articulation of its underlying science. It will surely be the same with psychology and neuroscience. At this level of complexity, intertheoretic reduction does not appear as the sudden takeover of one discipline by another; it more closely resembles a long and slowly maturing marriage.

References

Bartoshuk LM (1978) Gustatory System. In: Masterton RB (ed) Sensory integration. Plenum, New York, pp 503–567 (Handbook of behavioral neurobiology, vol 1)
Churchland PS (1986) Neurophilosophy: toward a unified understanding of the mind/brain. MIT Press, Cambridge, MA
Churchland PM (1989) A neurocomputational perspective: the nature of mind and the structure of science. MIT Press, Cambridge, MA
Churchland PM, Churchland PS (1990) Could a machine think? Sci Am 262: 32–37
Fodor JA (1975) The language of thought. Crowell, New York
Jackson F (1982) Epiphenomenal qualia. Philos Q 32: 127–136
Land E (1977) The retinex theory of color vision. Sci Am 237: 108–128
Nagel T (1974) What is it like to be a bat? Philos Rev 83: 435–450
Popper K, Eccles J (1978) The self and its brain. Springer, Berlin, Heidelberg, New York
Searle J (1980) Minds, brains, and programs. Behav Brain Sci 3: 417–457
Searle J (1990) Is the brain's mind a computer program? Sci Am 262: 26–31
Taylor C (1970) Mind-body identity: a side issue? In: Borst CV (ed) The mind/brain identity theory. Macmillan, Toronto, pp 231–241
Taylor C (1987) Overcoming epistemology. In: Baynes K, Bohman J, McCarthy T (eds) After philosophy: end or transformation? MIT Press, Cambridge, MA, pp 464–488

Language Comprehension in Ape and Child: Evolutionary Implications

E. S. Savage-Rumbaugh and *E. Rubert*

Summary

Previous investigations of the linguistic capacities of apes have focused upon the ape's ability to produce words with little concern for comprehension. By contrast, it is increasingly recognized that comprehension precedes production in the language development of normal human children (Golinkoff et al. 1987), and indeed may guide production. The study discussed in this chapter systematically compared the language comprehension skills of a 2-year-old child and a 9-year-old bonobo (*Pan paniscus)* by testing their comprehension of 660 novel sentences. Both subjects were exposed to spoken English and lexigrams from infancy and neither was trained to comprehend speech. All language acquisition was through observational learning. The results indicated that both subjects comprehended novel requests and simple syntactical devices. The bonobo decoded the syntactical devices of word order and recursion with higher accuracy than the child; however, the child performed better than the bonobo on the conjunctive, a structure which places a greater burden on short-term memory.

Man-Ape Relationship

"In trying to understand the functional principles governing the human nervous system, we must remind ourselves that our brain has evolved from earlier kinds of brains – that our brain was not built from scratch especially for us, but has capacities and limitations that are due to its historical origins ... The fundamental nature of cognition is rooted in the tricks by which assorted representational schema give organisms a competitive advantage in predicting" (Churchland 1986).

Apes are our closest living relatives (Sarich 1983). On the basis of biochemical data, it has recently been suggested that *Homo* and the African apes should share the genus nomenclature, *Pan* (Sibley and Ahlquist 1984, 1987). For the three species of Afrian apes – gorilla *(Gorilla gorilla)*, chimpanzee *(Pan troglodytes)*, and bonobo *(Pan paniscus)* – the currently available data indicate that both species of *Pan* share more DNA sequences with *Homo* than they do with *Gorilla* (Andrews and Martin 1987; Bishop and Friday 1986; Sibley and Ahlquist 1984, 1987). The estimated date of divergence between *Homo* and *Pan* is now placed at 4-6 million years (Sibley and Ahlquist 1987). Oldowan stone tools appear in the archeological record a bit later at 2-3 million years (Harris 1983).

The close biological connection between *Homo* and *Pan* has raised a variety of evolutionary questions, both behavioral and anatomical. The anatomical questions have focused upon the rapid evolution in brain size that began at the point of man-*Pan* divergence. Endocast studies reveal that the left temporal gyral area (in which language is often located in man) shows significant enlargement, relative to the right temporal area, around 2-3 million years ago (Falk 1983). Such observations suggest that perhaps some anatomical change, which made the appearance of proto-language possible, occurred at the same time. However, data from modern humans illustrate that language can be located in either hemisphere (Calvin and Ojemann 1980) and may also be present in hydrocephalics whose brains are much reduced in size and grossly abnormal in structure. Further complicating the picture is the clear evidence from direct electrical stimulation of the brains of epileptic patients (prior to undergoing surgery) that language can be localized in dramatically different areas in each brain (Calvin and Ojemann 1980).

Such localization findings suggest that language may be an acquired skill, developed and organized individually by each person through interactions with others. Many anthropologists (Noble and Davidson 1991) argue, on the basis of artifactual findings, that language is a very late invention, on the order of 30000 years ago, or long after *Homo sapiens* as we know them today had evolved.

Certainly, the behavioral differences between man and ape loom large as we look around the world today. Man has traveled through space to the moon and is in danger of destroying the very planet that gave rise to him, while the chimpanzee remains unchanged, getting food where it can with no home base, no apparent language, and no lasting structures of any sort. Many ask, how can these creatures be closely related to us unless there has been a qualitative reorganization from their brain to ours? However, evolution is conservative, using old parts to do things in slightly different ways, and the idea that a creature who shares 99 % of our DNA should be so different from ourselves is biological heresy.

In past times, man ascribed the differences between himself and other animals to the work of God. We were made in God's image and the animals were for our dominion. Such beliefs are now outmoded and have been replaced with the "brain reorganization" paradigm. Gross anatomical studies reveal more similarities than differences, aside from the dramatic increase in size (Passingham 1975). Whether God infused us with a soul that animals do not have or whether evolutionary pressure dramatically reorganized our brain, the effect is the same: a perspective of ourselves as irrevocably different from animals. The evidence for such a perspective is around us everywhere; there are cars, buildings, computers, governments, etc., to bolster our view. Yet we know that these things represent only a brief twinkle in man's history and that, for much of his past, he lived a life not very far removed from that of the ape. Gathering of food was very much a daily affair, shelters were temporary and reconstructed anew often; travel from place to place in search of new food sources was a fact of life. There were no city states and very few tools, little clothing.

In some places, this style of life continues to exist, forcing us to conclude that only a few behaviors set us clearly apart from animals. We have language, music, and art and apes seemingly do not. Of these skills it is language that enables us to plan for the future, discuss the past, and set up rules of economic trade. Hence it is language that has become the point of extreme focus in all discussions of what it is to be human and in all attempts to understand how it is that our brain functions.

What kinds of evolutionary forces would conspire to bring about the appearance of a skill so novel and complex as language? Out of what old parts did nature fashion this new and powerful device? When we strip away modern technology and culture and look at ourselves as we lived even 200000 thousand years ago – arguably *Homo sapiens* in anatomy, but with few tools, extremely short life spans, and a nomadic existence – the reorganization of our brain seems less dramatic.

Looking back in time seems to separate us less from apes, but what of looking at apes more closely? Could we have overlooked some things that may be indicative of greater similarity to ourselves than we realize? Indeed what would man be like if his brain organization were left intact but shrunk by two-thirds in size, without a concomitant decrease in body mass? One might suspect that we could do all of the things that we do now, but just not as well. But what does "just not as well" mean? Might we do some things so poorly that we would not recognize the behaviors at all? If our motor coordination were not as good, would we still dance? Would the rhythmic movements we were still able to make be called "dancing?" Certainly dancing is something that other animals have difficulty doing with anything like the grace or complexity demonstrated by man. What if our ability to control the coordination of our tongue and our larynx was not as good; would we still "talk?"

The point is that some of the things we view as uniquely human may be better understood as the product of a large brain, rather than those of a reorganized brain. A computer with a large memory buffer appears to do more things at the same time than one with a small buffer, but both operate upon information in the same way. Similarly, it may be that our brain functions in a manner very much like that of the ape, but that ours is able to keep track of information along a greater number of independent parallel tracts. This capacity generates a processing ability which leads us to a more complete understanding of relationships between events in the world around us. It also allows us to plan and coordinate everything more precisely, from the motor movement required in dance steps to those required by speech.

If this is the case, we should expect to find apes doing many of the things that we do, but less adeptly. Concomitantly, we should expect monkeys to be doing many of the things that apes do, but less adeptly. Strong support for this perspective comes from the capacities exhibited may apes exposed to human language from birth. Such experience permits the ape to decode human speech sounds, to associate meaning with those sounds, and to understand novel sentences (Savage-Rumbaugh et al. 1986; Savage-Rumbaugh 1988).

To accomplish such things, an ape must identify rapidly sequenced phonemic units that it cannot produce. It must differentiate word boundaries, boundaries that are rarely clear in natural speech and that are difficult for people to identify in non-native languages. It must then analyze the situations surrounding the occurrence of different sound units to determine how they are used by different speakers at different times, to arrive at a reasonable approximation of the "potential range of meanings of each unit." Having done this, the ape will still need to analyze each new linguistic situation and to determine how the units are being used at that moment with other units. When syntactical cues such as word order or phrase markers are employed, the ape will need to understand the way in which these devices function and to apply them appropriately, regardless of the specific vocabulary units in the utterance. These are extremely complex skills, indeed so complex that many have been tempted to conclude that the only way in which a human child could learn to do these things is with the assistance of an innate hypothetical mechanism termed the language acquisition device (LAD; Chomsky 1988).

Testing Language Comprehension in Two Species: Method

The English comprehension skills of a 9-year-old bonobo (*Pan paniscus*) were systematically compared with those of 2-year-old child. Unlike previous ape-child comparisions, which used an "idealized child" (formed from the data of various investigators using different methods), this study sought to test the capacities of a normal human child (Alia) and an ape (Kanzi) in the same manner. Moreover, the rearing histories of both subjects were known to be sufficiently similar to make the comparison valid (for a detailed report of this work and the entire corpus of data see Savage-Rumbaugh, Murphy, Sevcik and Rumbaugh in press).

Both subjects were exposed to a similar linguistic environment from birth, in that neither was "trained" to talk. Both had similar experiences with lexigrams (e.g., geometric symbols that serve as words in our laboratory). Additionally, both subjects shared a primary caregiver, Jeannine Murphy, who was the mother of the child and whose language input and caretaking behavior were similar for both subjects.

The subjects were presented with the same 660 novel sentences to determine the extent to which they comprehended the semantic and syntactic aspects of language. The sentences were presented in spoken English in a normal voice. Relative clauses were utilized, as well as word order reversals, to investigate comprehension of these specific syntactical devices.

At first, the experimenter simply sat in front of the subject and, while playing with toys, asked him or her to do something in a normal manner, as part of the ongoing play sessions. After 244 trials (each entailing a different request) the subjects understood the "game" and blind controls were instituted. The experimenter sat behind a one-way mirror and spoke to the subject in a normal voice, asking him or her to carry out a specific action with the objects in front of him or her, or to go to another location and do something.

The sentences fell into seven different semantic groups, and within three of these groups there were different syntactical subtypes. All sentence types, along with examples of each type, are presented in Table 1.

Table 1

Sentence Type 1-A. Put object X in/on transportable object Y.
- Put the rubber band on your ball.
- Put the sparklers in the shoe.
- Put the telephone on the TV.
- Put the raisins in the yogurt.
- Put the rubber band on the milk.

Sentence Type 1-B. Take (put) object X in/on nontransportable object Y.
- Put the shoe in the potty.
- Take your ball to the table.
- Take the phone to the colony room.
- Take the banana outdoors.
- Take the tomato to the micro-wave.

Sentence Type 2-A. Give (Show/get) object X to animate A.
- Go get a coke for Rose.
- Give the knife to Kelly.
- Show me the can opener.
- Give the cereal to Panbanisha.
- Give the doggie some milk.

Sentence Type 2-B. Give (show) X and Y to animate A.
- Give the dog and the shot to Kelly.
- Give the apple and the hat to Rose.
- Show Sue the toothpaste and the milk.
- Show me the ball and the rubberband.
- Give me the lighter and the water.

Sentence Type 2-C. (Do) action A on animate A.
- Give Rose a hug.
- Go vacuum Liz.
- Groom the doggie.
- Go scare Matata.
- Hammer the doggie.

Sentence Type 2-D. (Do) action A on animate A with object X.
- Tickle Rose with the bunny.
- Go put some soap on Liz.
- Put the monster masks on Linda.
- Tickle Liz with the umbrella.
- Can you put the bunny on your hand?

Sentence Type 3. Do action A on object A (with object B).
- Vacuum your ball.
- Bite the stick.
- Knife your ball.
- Hammer the vacuum.
- Hit your ball with the sugar cane.

Sentence Type 4. Announce information.
- Kanzi is going to tickle Liz with the bunny.
- There is a new ball hiding at Sherman and Austin.
- There is a surprise hiding in Matata's yard.
- I want Kanzi to grab Rose.
- Linda is going to chase Kanzi.

Table 1. Continued

Sentence Type 5-A. Take object X to location Y.
Take the doggie to the colony room.
Take the toothpaste outdoors.
Take the hat to the bedroom.
Take the mushrooms to the T-room.
Take the tomato to the micro-wave.
Sentence Type 5-B. Go to location Y and get object X.
Go outdoors and get a banana.
Go to the potty and get sparklers.
Go to the micro-wave and get a shoe.
Go to the colony room and get a ball.
Go to the bedroom and get a shot.
Sentence Type 5-C. Go get the X that's in location Y.
Go get the melon that's in the T-room.
Go get the lettuce that's in the micro-wave.
Go get the noodles that are in the bedroom.
Go get the snake that's outdoors.
Go get the collar that's in the refrigerator.
Sentence Type 6. Make pretend animate A do action A.
Can you make the bunny eat the sweet potato.
Make the toy orang bite Rose.
Make the doggie bite the snake.
Can you make the doggie chase the bug?
Make the snake bite Linda.
Sentence Type 7. All remaining sentences
Push the knife under the door.
Take the potato outdoors and get the apple.
Take Rose to the refrigerator and get some food.
Open the Jello and pour it in the juice.
Tell Rose that you want to go outdoors.

Results

Kanzi's and Alia's total scores, and their scores broken down by sentence type, are shown in Table 2. Overall, Kanzi was correct on 72 % of all sentences and 74 % of the blind sentences. Alia was correct on 67 % of all sentences and 66 % of the blind sentences. There were no significant differences between the blind and nonblind conditions in terms of overall performance for either subject. The overall high performance of both subjects is strong evidence of their ability to comprehend most sentence types and subtypes.

Kanzi performed significantly better on some sentence types than others, both for the overall data set χ^2 (24, N = 660) = 57.17, p <.0002), and in the blind condition, χ^2 (24, N = 415) = 54.88, p. <0003) as did Alia, both for the overall data set χ^2 (24, N = 589) = 67.02, p. <0000 and in the blind condition χ^2 (24, N = 409) = 68.66, p. <0000. On blind trials, Kanzi and Alia differed significantly from one another on sentence type 2-B, χ^2 (2, N = 19) = 9.00, p.<.01, sentence type 3.χ^2 (2, N = 49) = 5.76, p. <05, sentence type 5-B, χ^2 (2, N = 38) = 7.67, p.

Table 2. Performance by sentence type

Sentence Type	Number of trials		Nonblind condition		Blind condition	
	Total	Blind	Kanzi	Alia	Kanzi	Alia
All[c]	660	416	69 %	68 %	74 %	66 %
1-A	129	63	64 %	70 %	66 %	75 %
1-B	50	17	72 %	74 %	77 %	71 %
2-A	69	46	87 %	73 %	78 %	86 %
2B[a]	21	19	0 %	0 %	26 %	57 %
2-C	19	12	83 %	75 %	83 %	91 %
2-D[b]	86	49	65 %	54 %	78 %	57 %
3[b]	83	49	52 %	65 %	84 %	62 %
4	16	12	75 %	100 %	58 %	75 %
5-A	85	58	70 %	62 %	78 %	72 %
5-B[b]	46	38	100 %	100 %	79 %	50 %
5-C[b]	35	35	NG	NG	77 %	50 %
6	10	9	100 %	100 %	67 %	44 %
7[b]	11	9	0 %	0 %	100 %	44 %

[a] Alia performed significantly better than Kanzi in the blind condition.
[b] Kanzi performed significantly better than Alia in the blind condition.
[c] This is the total number of sentences presented to Kanzi. Alia may have received a few less in some cases (see text).

<02, and sentence type 7, χ^2 (2, $N = 9$) = 6.92, p. <.03, with Kanzi doing better than Alia on all sentence types except 2-B.

Both subjects were capable of processing the semantic and syntactic information in the sentences presented to them. Moreover, the manner in which they did so revealed they did not interpret the words in sentences as randomly juxtaposed events, to be acted upon independently. Instead, they inevitably attempted to carry out a complex set of related actions which reflected their interpretation of the semantic and syntactic features of each novel utterance. For example, Kanzi's solution to "Put the water on the carrot" was to toss the carrot out into the rain. Such innovative actions revealed a sophisticated processing of the speaker's intent (in this case, to get the carrot wet), rather than a rote unthinking solution. Even when the subjects failed, they virtually never did so in a way which would suggest that they were randomly tossing together a word salad.

Both subjects also responded appropriately to unusual sentences. For example, Kanzi correctly carried out the sentence, "Feed your ball some tomato" (see Fig. 1).

Since the word "feed" was never juxtaposed (in Kanzi's prior experience) with the word "ball," his appropriate response indicates that he understood the action encoded in the verb "feed" was to be directed toward a "ball," regardless of the plausibility of the request.

Analysis of the types of errors indicated that both subjects were far more likely to make semantic that syntactic errors. For example, when presented with 1-A sentences (Put X in Y), it was possible to err by performing the inverse action (i.e., putting Y in X). Surprisingly however, such errors of inversion were rare:

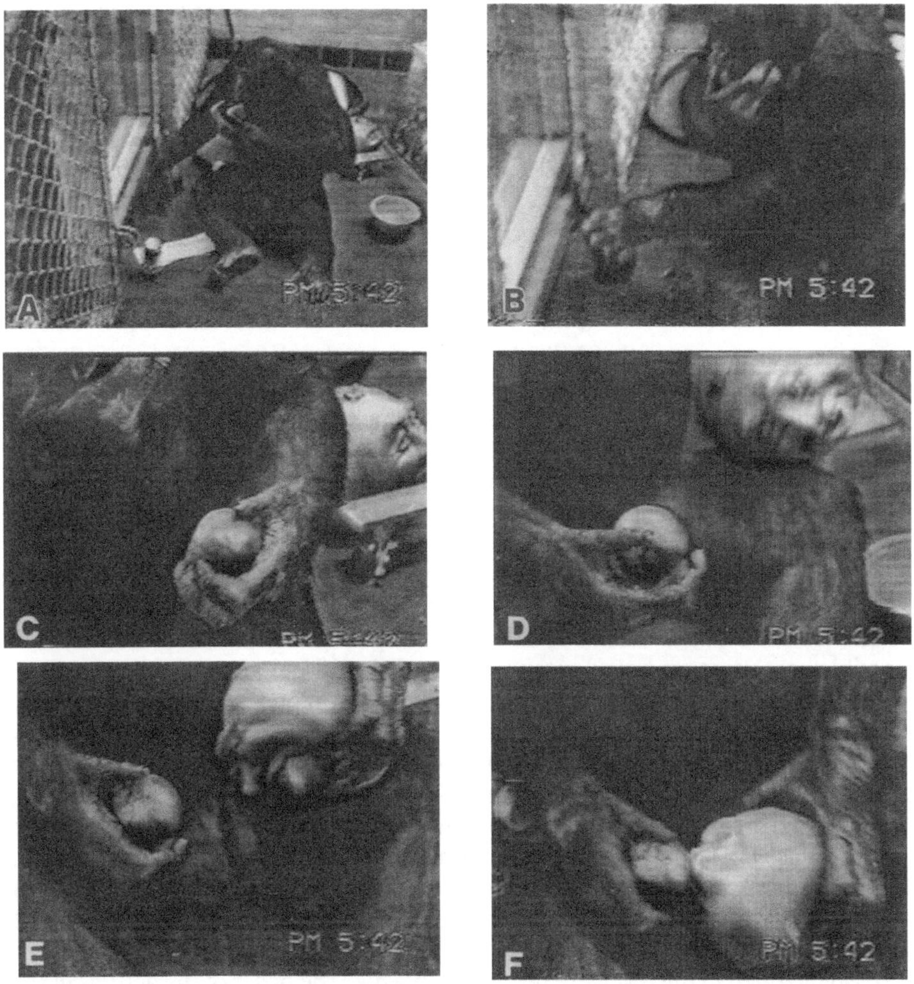

Fig. 1 A–F. "Feed your ball some tomato."
A Kanzi listens to the sentence "Feed your ball some tomato." *B* Kanzi selects the tomato, *C, D* Kanzi shifts the tomato between his hands while looking for the "ball". The "ball" is a soft sponge ball which has a face embedded within in, *E* Kanzi picks up the ball and orients the "mouth" on the ball toward the tomato, *F* Kanzi places the tomato in the ball's mouth

Kanzi made only five and Alia made only six. If the subjects had been responding to 1-A sentences without regard for word order, inversions should have been much higher, as chance responding would have resulted in approximately 60 inversions. By contrast, semantic errors (i.e., those in which the subjects interpreted the verb or object term inappropriately) were common: Kanzi made 35 semantic errors while Alia made 21. The high frequency of semantic errors occurred in spite of the fact that, with only one or two exceptions, these sentences were composed of individual words that both subjects could readily identify in a naming task; hence their semantic errors could not be attributed to the lack of an appropriate vocabulary.

These findings go against the general view that semantic content is used to bootstrap the linguistic system (Pinker 1987 b). They are also at odds with the commonly accepted tenet that human subjects possess a language acquisition device (LAD) or unique ability to perform parsing operations (Chomsky 1965; Lenneberg 1967).

Reversible Sentences

For a more direct test of the ability to interpret word order, some sentences were presented which required actual reversals, rather than potential reversals. Sixty-eight sentences, forming 34 pairs of such sentences, are presented in Table 3. They are of two types: Group 1 sentences are reversible with the same verb, whereas Group II sentences are reversible, but the verb changes from "take" to "go get." Kanzi was correct on 80 % of these sentences and he correctly executed both sentences of a given pair for 65 % of the pairs. Alia was correct on 64 % of the sentences and on 33 % of the pairs. When the reversible sentences with a common verb are considered, Kanzi was correct on 29 of the 40 possible sentences and on both sentences in 11 of the 20 pairs. Alia was correct on 25 of 38 possible sentences on both sentences in 6 of the 18 pairs.

Overall, on the reversible sentences with the common verb, Kanzi made only three errors of inversion and Alia made only four, indicating that *even when they were incorrect on reversal sentences, their errors usually were not generated by an inability to interpret word order,* but instead were semantic errors. For example, when Kanzi heard "Pour the milk in the cereal" he poured the milk in the mushrooms. That is, he initiated an action using the milk (rather than the cereal) but poured the milk onto the wrong item. Similarly, when Alia was asked to put the peaches in the tomatoes, she responded by putting the peaches in the yogurt.

Also included in the test were sentences in which the order of the actions in the response should have varied, even though the order of the location and object terms did not vary (i.e., "Put the melon in the potty" versus "Get the melon that's in the potty.") Thus the test presented the subjects with a double challenge. On the one hand, some sentences required that the order of X and Y be treated as a signal about the sequence that ensuing actions should take place (linear sentences). In other cases, the order of X and Y was to be ignored (embedded sentences). Both subjects were able to respond to word order cues in the linear sentences, and to ignore word order in the embedded sentences, though Kanzi did significantly better than Alia on these sentence types (Figs. 2, 3).

For a subset of the embedded sentences (Go get the X that's in Y), the object to be retrieved was located in a group of objects in front of the subjects, as well as at another location. Embedded sentences were then contrasted with linear sentences. If the subjects did not understand the nature of the recursive structure, they would most probably respond to both the linear and the

Table 3. Reversible sentences presented in database

Kanzi	Alia	Group 1: Fully reversible sentences
C	NG[a]	Can you put some oil on your ball?[a]
C	C	Put the ball in the oil.
PC	C	Put the hat on your ball.
I	I	Put the ball on the hat.
C	C	Put the ball on the rock.[a]
C	NG	Can you put the rock on your ball.
C	C	Put the pine needles in your ball?
C	W	Can you put the ball on the pine needles?
C	C	Put some water on your carrot.
C	C	Put your carrot in the water.
PC	C	Pour the milk in the cereal.
C	I	Pour the cereal in the milk.
C	PC	Pour the coke in the lemonade.
C	PC	Pour the lemonade in the coke.
C	C	Pour the juice in the egg.
C	C	Put the egg in the juice.
C	C	Put the rock in the water.
C	C	Put the water on the rock.
C	C	Put the raisins in the water.
C	I	Pour the water on the raisins.
PC	PC	Put the melon (peaches) in the tomatoes.
C	C	Put the tomatoes in the melon (peaches.)
C	C	Put the milk in the water.
C	PC	Pour the Perrier water in the milk.
C	C	Put the tomato in the oil.
C	C	Put the oil in the tomato.
I	PC	Put the shoe in the raisins.
C	C	Put the raisins in the shoe.
C	C	Pour the juice in the Jello.
C	C	Open the Jello and pour it in the juice.
C	C	Rose (Nat) is gonna chase Kanzi (Alia).
PC	I	Kanzi (Alia) is going to chase Rose (Mom).
C	C	Liz (Linda) is going to tickle Kanzi (Alia).
PC	I	Kanzi (Alia) is gonna tickle Liz (Linda).

Table 3. Continued

Kanzi	Alia	Group 1: Fully reversible sentences
C	C	Kanzi (Alia) is going to tickle Liz (Nat).
PC	C	Liz (Nat) is going to tickle Kanzi (Alia).
C	C	Kanzi (Alia) is going to tickle Liz (Nat) with the bunny.
PC	C	Liz (Nat) is going to tickle Kanzi (Alia) with the bunny.
C	PC	Make the doggie bite the snake.
C	C	Make the snake bite the doggie.
Total		Kanzi 31/40 correct Alia 26/38 correct
Kanzi	Alia	Group 2 sentences; reversals which take different verbs (take or put versus go... get)
PC	C	Take the carrots outdoors.[a]
C	C	Go outdoors and find the carrot.
W	PC	Take the melon (pears) to the refrigerator.
C	PC	Go to the refrigerator and get the melon.
C	C	Take the potato to the bedroom.
C	C	Go to the bedroom and get the potato.
C	NG	Could you take the pine needles outdoors?[a]
C	C	Go outdoors and get the pineneedles.
C	C	Take the orange outdoors.
C	C	Go outdoors and get the orange.
C	C	Take the umbrella (box) outdoors.
C	C	Go outdoors and get the umbrella (box).
C	C	Take the pineapple (apple) outdoors.
C	W	Go outdoors and get the pineapple (apple).
C	C	Take the lighter (matches) outdoors.[a]
C	NR	Go outdoors and get the lighter (matches).
OE	PC	Take the stick to the bedroom.
C	NR	Go to the bedroom and get the stick.
C	PC	Take the tomato to the bedroom.
C	W	Go to the bedroom and get the tomato.
C	C	Take you collar (watch) to the bedroom.
C	NR	Go to the bedroom and get the collar (watch).
C	OE	Take the raisins to the bedroom.
C	PC	Go to the bedroom and get the raisins.
C	C	Take the ice back to the refrigerator.
C	C	Go to the refrigerator and get some ice.
C	C	Take the potato outdoors.
C	C	Go outdoors and get the potato.
Total		Kanzi 25/28 correct Alia 16/28 correct

Table 3. Continued

Kanzi	Alia	Group 1: Fully reversible sentences
Kanzi	Alia	Group 3 sentences: word order is constant but action is reversed
PC	OE	Take the rock outdoors.[a]
C	C	Go get the rock that's outdoors.
C	C	Take the stick outdoors.
OE	C	Go get the stick that's outdoors.
C	C	Take the snake (bug) outdoors.
C	C	Go get the snake (bug) that's outdoors.
C	C	Take the banana outdoors.
C	C	Go get the banana that's outdoors.
C	W	Take the tomato to the microwave (oven).
C	PC	Go get the tomato that's in the microwave (oven).
C	C	Put the raisins in the refrigerator.
C	PC	Go get the raisins that are in the refrigerator.
C	C	Put your collar (watch) in the refrigerator.
C	C	Go get you collar (watch) that's in the refrigerator.
C	C	Put your apple in the micro-wave.
C	OE	Go get the apple that's in the oven.
C	C	Put the melon (peaches) in the potty.
C	M	Get the melon (peaches) that's in the potty.
C	C	Put the sparklers in the potty.[a]
PC	OE	Go to the potty and get the sparklers.
C	C	Put the doggie in the refrigerator.
W	W	Go get the doggie that's in the refrigerator.
Total		Kanzi 18/24 correct Alia 14/24 correct

[a] NG, this sentence was not given to Alia.
[c] subject responded correctly to the sentence; PC, subject responded correctly to one or more nouns or verbs in the sentence; OE, subject responded correctly to the sentence but also added something extra for example, when asked to retrieve object X, the subject also brought back additional items; W, subject responded incorrectly to the sentence; NR, subject did not respond to the sentence at all

embedded sentences in the same way. While Alia's data were somewhat ambiguous, Kanzi's data indicated that he processed both sentence types. He reacted as though the linear structures were ambiguous, which they were, and to the recursive structures as though the embedded phrases modified the object, which they did. That is, when asked to "Go to Y and get X" he sometimes retrieved the X object in front of him, as well as the one in location Y. However, when asked to "Get the X that's in Y", he virtually always retrieved the X object that was in the other location, and ignored the X object immediately in front of him.

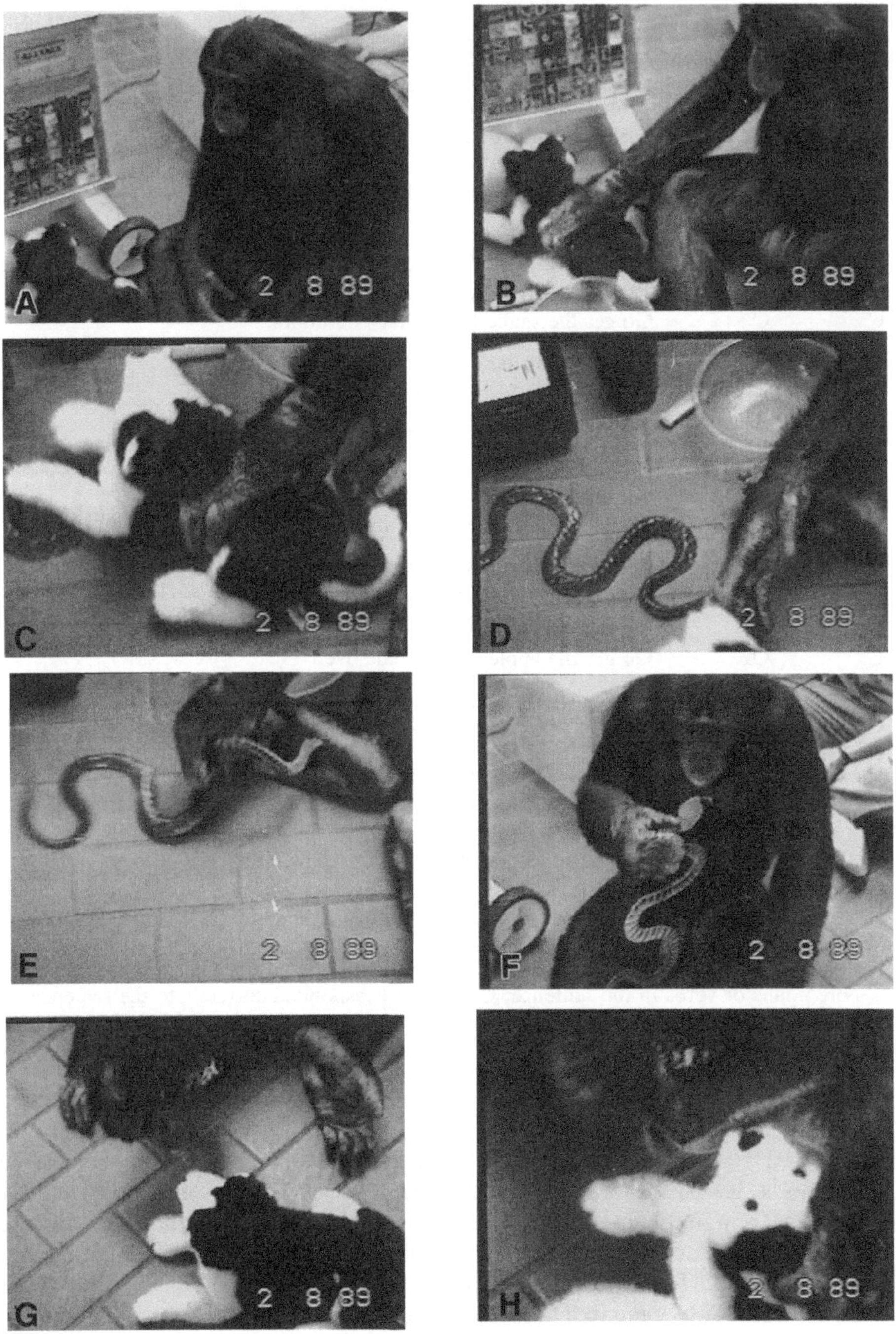

Fig. 2 A–H. Make the snake bite the doggie.
A Kanzi listens to the sentence "Make the doggie bite the snake", B Kanzi picks up the doggie, C Kanzi puts the doggie down by the snake, D Kanzi puts his finger in the doggie's mouth, E Kanzi picks up the snake, F Kanzi looks at the snake's head, G Kanzi moves the snake toward the dog's mouth, H Kanzi opens the dog's mouth and inserts the snake's head

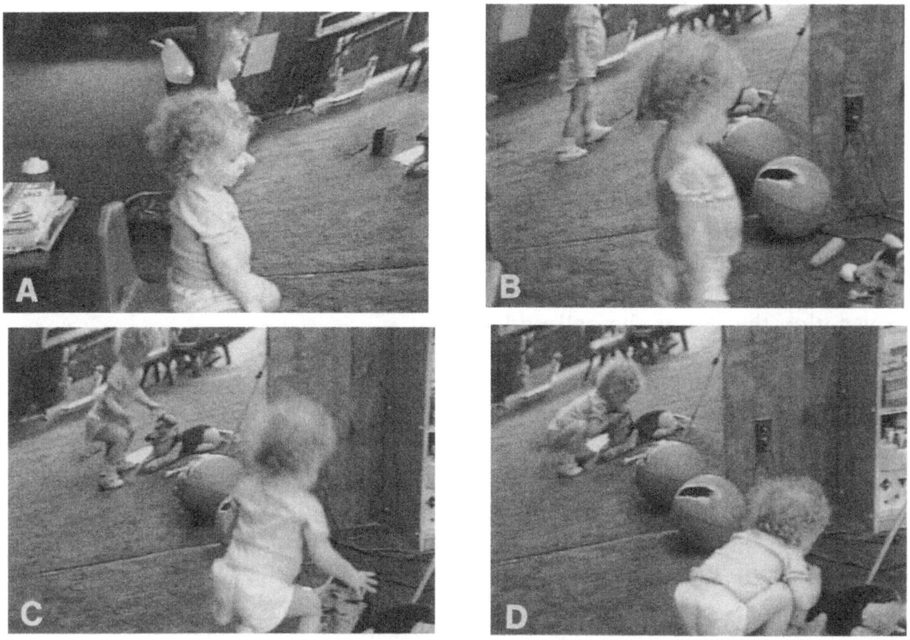

Fig. 3 A–D. Make the doggie bite the snake.
A Alia listens to the sentence "Make the snake bite the doggie", B Alia approaches the dog, C Alia bends over by the dog, D Alia bites the dog

When there was a significant difference between Kanzi and Alia, it was Kanzi who performed better than Alia in all cases but one, type 2-B sentences. The 2-B sentences required the subjects to give the experimenter two objects. Kanzi's error was that he typically acted as if he had heard the name of only one object. Sometimes he selected the object mentioned first, other times he selected the object mentioned last. When Kanzi was reminded of the second object by the experimenter, he always quickly found it and handed it to the experimenter as though he knew what to do with it. Alia, by contrast, nearly always selected both objects. Moreover, she picked up both before attempting to give either one to the experimenter.

The construction of 2-B sentences was linear and quite simple; there was no reason to believe that Kanzi could not understand it. Indeed, on some trials he selected two objects. Nonetheless, 2-B sentences were the most difficult type of all for Kanzi. Because he understood both the words and the general sentence structure, it appears that Kanzi's difficulties with 2-B sentences were a function of short-term memory limitations. There was no semantic or syntactic link between the items he was to give; he simply had to remember both items. By contrast, in a sentence such as "Put the rock in the water," a relationship was constructed between rock and water that intergrated the two. Thus, syntactical relationships, rather than proving difficult for Kanzi, appeared to make it easier for him to chunk information and consequently to remember what he was supposed to do.

Language Acquisition Device

Certainly no one has suggested that chimpanzees have a language acquisition device; nonetheless, it is clear that the language comprehension skills of apes reach at least the same level as those of a two-year-old child. However, early experience appears to be critical to the appearance of such skills, as though the ape's brain, if exposed to certain types of stimuli, goes about organizing its own language acquisition device in response (Savage-Rumbaugh et al. 1985, 1989, 1991). The critical event appears to be exposure to a spoken English in a communicative environment, where the ape is spoken to as a child, with the expectation that he or she will understand.

Multisymbol Utterances

Surprisingly, under these minimal conditions, apes have learned to understand human language to a rather complex degree and to make simple needs known, even though they cannot produce the sounds of the English language. If their vocal tract permitted them to produce sounds similar to our own, it is reasonable

Table 4. Distribution of two-element semantic relations in Kanzi's corpus.

Relation	N	Example (of dominant order)
Action-action	92	*Tickle bite,* then positions himself for researcher/caregiver to tickle and bite him.
Action-agent	120	*Carry person (gesture),* gesturing to Phil, who agrees to carry Kanzi.
Agent-action	13	
Action-object	39	*Keepaway balloon,* wanting to tease Bill with a balloon and start a fight.
Object-action	15	
Object-agent	7	*Balloon person (gesture),* Kanzi gestures to Liz; Liz gives Kanzi a balloon.
Agent-object	1	
Entity-demonstrative	182	*Peanut that (gesture),* points to peanuts in cooler.
Demonstrative-entity	67	
Goal-action	46	*Coke chase,* then researcher chases Kanzi to place in woods where coke is kept.
Action-goal	10	
Entity-entity	25	*M&M grape,* caregiver/researcher: "You want both of these foods?" Kanzi vocalizes and holds out his hand.
Location-location	7	*Sue's-office childside,* wanted to go to those two places.
Location-entity	19	*Playyard austin,* wants to visit Austin in the playyard.
Entity-location	13	
Entity-attribute	12	*Food blackberry,* after eating blackberries, to request more.
Attribute-entity	10	
Miscellaneous relations	37	These include low frequency (less than seven) such as attribute of action, attribute of location, affirmation, negation, and those involving an instrument.

to conclude they would be able to convey many of the things that 2- and 3-year-old children communicate. However, with their limitations on short-term memory, it is not likely that apes would be able to convey them in the same way or with as much detail or elaboration. They tend to make simple, two-word constructions, they follow English word order and are even able to invent simple syntactical rules of their own (Greenfield and Savage-Rumbaugh 1990). Table 4 summarizes a corpus of Kanzi's multiword utterances, collected at 5 1/2 years of age. It can be seen that most of Kanzi's combinations were novel and that is was the exception, rather than the rule, for him to repeat himself.

Unlike Nim (a chimpanzee who learned signs), Kanzi did not tend to imitate the utterances of others (Greenfield and Savage-Rumbaugh 1990). Less that 3 % of all his utterances could be classified as imitation. When Kanzi produced multiword utterances, the additional words did not function as "wild-cards" or repetitions, neither did they state the obvious, as did Nim's multiword utterances (see Savage-Rumbaugh and Brakke (1990) for a detailed analysis of Kanzi's multiword utterances).

Language Acquisition Schemas

Symbol acquisition in the chimpanzee appears to begin with the learning of routines. These are not planned routines, but rather are routines which emerge out of everyday life. The chimpanzee's daily interactions with caretakers, although not experimentally programmed, can be viewed as a series of interindividual "routines", which become ever more complex and interchangeable with maturation and experience. The word "routine" is used to mean a more or less regularly sequenced set of interindividual interactions that occur in a relatively similar manner across time, or at different times. The sequence of interactions may vary, as may the words used in connection with the interaction, however, each routine is carried out for a specific purpose.

Examples of routines include changing diapers, getting ready to go outdoors, taking a bath, riding in the car, looking at a book, blowing bubbles, putting items in the backpack, visiting with other apes, playing a game of tickle, and traveling down various trails in the forest. The chimpanzee may be a willing or unwilling participant within such routines. Similarly, he/she may play different roles on different occasions, as participant or as observer.

The learning of such routines occurs regardless of whether the chimpanzee actively seeks to carry out the routine or finds it reinforcing. Human caretakers invevitably accompany most parts of a routine with gestural, lexical, and vocal markers directed to the ape. The caretakers are not instructed to do so by experimental design, but rather do so as a natural extension of their tendency to behave in this fashion with young children. Take, for example, a sample routine: "blowing bubbles." This routine has the following components:

a) finding the bubbles,
b) opening the bubbles,
c) getting the bubble-wand out of the bubble jar,
d) blowing bubbles, and
e) watching or attempting to pop the bubbles.

Both participants may take turns at any of these activities. Once the routine is understood, it will be initiated by the ape. At first, such initiations will be rather primitive in the sense that they are action based and context dependent. For example, the chimpanzee may see the bottle of bubbles among other toys and will pick it up and look at the caretaker. By selecting the bubbles from among other things, the chimpanzee has thus conveyed its desire to execute the "bubble-blowing" routine. Later it may simply point to the bubbles and look at the caretaker. Still later, it will point to the "bubbles" lexigram and turn to the caretaker.

In so doing, the ape has moved from being a passive observer of a routine, to an active participant, to a primitive initiator, to a communicator symbolically announcing his/her intentions to another party. The process occurs very naturally, without the caretaker intentionally or knowingly structuring the transition from passive receptive comprehension to productive knowledge and use. It appears to happen most rapidly with routines that are clearly structured and effectively marked.

The process is driven not by the caretaker, but by the need to *coordinate* interindividual interactions within routines. The verbal and nonverbal segmental markers allow such coordination to take place. It is because such markers allow the recipient to predict what is going to happen next that they are learned. Thus, although the ape's caretaker may serve as a "scaffolder" (Bruner 1983) or as a catalyst (Nelson 1985), the caretaker is not the "driving force" behind language acquisition in the ape. Rather it is the desire of the ape to be able to accurately predict what is going to happen to it next. To the extent that the caretaker's markers serve to signal this, the markers are learned receptively because they serve to make the world more predictable.

Conclusion

Therefore, language as a representational schema, gives those who understand it, the ability to predict the future actions of others and, in so doing, to come to react to or alter such actions before they occur. The ability to predict observable events, such as whether or not the predator one sees will approach, is very frequent in the animal kingdom. Such predictions are based upon observable events, rather than upon inferences regarding "intent." However, in man, the ability to determine the "intent" of conspecifics from their communications has become a skill of paramount importance. By knowing what it is that others are going to do in the future and in different places, we can coordinate our behavior with theirs. The fact that chimpanzees can comprehend symbolically communicated intentions, at least as accurately as 2-year-old children, suggests that they may well be using vocalizations to convey novel information regarding their future intent in the wild. Since studies of naturally occurring vocalizations have looked only at what other animals do immediately after they hear a sound, we have perhaps underestimated the complexity of communication in higher mammals. Kanzi's ability to comprehend syntactically and semantically novel sentences suggests that we have been too quick to conclude that such creatures have "inflexible communication systems."

Language is not likely to be dependent upon a capacity to produce a wide range of sounds; if so, parrots would be conversing with us, rather than vocalizing at us. Language is instead dependent upon the ability to make complex inferences, regarding the intent of speakers on the basis of verbal information alone. To do so, it is necessary to take into account a wide variety of situational factors and the relationships among them, for the intent behind the utterance of a word or group of words is rarely precisely the same on different occasions. The larger the brain, the more relevant information can be evaluated in the assessment of the nature of the speaker's signal and the judgment of his or her future intent, and consequently, the greater the ability to predict and coordinate future inter-individual behavior.

Acknowledgements. The research described in this paper and its preparation were supported by National Institutes of Health grant NICHD-06016 which supports the Language Research Center, cooperatively operated by Georgia State University and the Yerkes Regional Primate Research Center, Emory University. The research was supported in part by RR-00165 to the Yerkes Regional Primate Research Center, Emory University.

Special appreciation is extended to Kelly McDonald without whose assistance this work would never have been possible. She participated in all phases of rearing and testing Kanzi. Thanks are also extended to Mike Tomasello and to Ian Davidson for most helpful comments on earlier drafts of this paper.

References

Andrews P, Martin L (1987) Cladistic relationships of extant and fossil hominoids. J Human Evolution 16: 101–108

Bishop MJ, Friday AE (1986) Molecular sequences and hominoid phylogeny. In: Wood B, Martin L, Andrews P (eds) Major topics in human evolution. Cambridge, University of Cambridge Press, pp 150–156

Brunner J (1983) Child's talk: learning to use language. Norton, New York

Calvin WH, Ojemann GA (1980) Inside the brain. New American Library, New York

Chomsky N (1965) Aspects of a theory of syntax. MIT Press, Cambridge

Chomsky N (1988) Language and problems of knowledge. The Managua Lectures. MIT Press, Cambridge

Churchland PS (1986) Neurophilosophy: Toward a unified science of the mind/brain. MIT Press, Cambridge

Falk D (1983) Cerebral cortices of East African early Hominids. Science 221: 1072–1074

Golinkoff RM, Hirsh-Pasek K, Cauley KM, Gorden L (1987) The eyes have it: lexical and syntactic comprehension in a new paradigm. J Child Language *14*: 23–45

Greenfield P, Savage-Rumbaugh S (1990) Imitation, grammatical development, and the invention of protogrammer by an ape. In: Krasnegor N, Rumbaugh DM, Studdert-Kennedy M, Scheifelbusch (eds) Biobehavioral foundations of language development. Erlbaum, Hillsdale

Harris JWK (1983) Cultural beginnings: Plio-Pleistocene archaeological occurrences from the Afar, Ethiopia. African Archaeological Rev 1: 3–31

Lenneberg E (1967) Biological foundations of language. Wiley, New York

Nelson K (1985) Making sense: The acquisition of shared meaning. Academic, New York

Noble W, Davidson I (1991) the evolution of modern human behavior: Talking about its archaeological record. Man, in press

Passingham RE (1975) Changes in the size and organization of the brain in man and his ancestors. Brain, Behav Evolution 7: 337–359

Pinker S (1987) Constraint satisfaction networks as implementations of nativist theories of language acquisition. In: MacWhinney B (ed) Mechanisms of language learning. Erlbaum, Hillsdale

Sarich VM (1983) Retrospective on hominoid macromolecular systematics. In: Ciochon RL, Corruccini RS (eds) New interpretations of ape and human ancestry. Plenum, New York, pp 136–150

Savage-Rumbaugh ES, Brakke K (1990) Animal language: Methodological and interpretive issues. In: Beckoff M, Jamieson D (eds) Interpretation and explanation in the study of animal behavior: Vol 1. Interpretation, intentionality, and communication. Westview, Boulder

Savage-Rumbaugh ES, Sevcik R, Rumbaugh DM, Rubert E (1985) The capacity of animals to acquire language: Do species differences have anything to say to us? Phil Trans R Soc London, B 308: 177–185

Savage-Rumbaugh ES, Romski MA, Hopkins WD, Sevcik R (1989) Symbol acquisition and use by *Pan troglodytes, Pan paniscus,* and *Homo sapiens.* In: Heltne PG, Marquardt LA (eds) Understanding chimpanzee. Harvard University Press, Cambridge

Sibley CG, Ahlquist JE (1984) The phylogeny of hominoid primates as indicated by DNA-DMA hybridization. J Mol Evolution 20: 2–15

Sibley CG, Ahlquist JE (1987) DMA hybridization evidence of hominoid phylogeny: Results from an expanded data set. J Mol Evolution 26: 99–121

A Cognitive Neuroscience of Alzheimer's Disease: What Can Be Learned from Studies of visual Imagery?

S. M. Kosslyn and *I. E. Dror*

Summary

Complex cognitive abilities are subserved by sets of component subsystems working together. Behavioral deficits, such as those that occur in Alzheimer's disease, arise when subsystems malfunction or interact improperly. A theory of component subsystems of visual mental imagery is described, which specifies five major sets of components; these components are hypothesized to be implemented in different regions of the brain. The theory is used to guide investigations of the effects of normal aging on different aspects of imagery, and the results of these studies suggest that normal aging selectively affects aspects of imagery that depend on frontal lobe function. These findings are used to formulate hypotheses about possible relations between Alzheimer's disease, normal aging, and visual mental imagery.

Introduction

Cognitive functions break down selectively following a brain disorder, whether due to a focal injury such as a stroke, or more diffuse degeneration, such as Alzheimer's disease. The precise way functions break down depends on the structure of normal cognitive mechanisms. In this chapter we briefly consider a theory of the major components of visual mental imagery, and consider the implications of this architecture for understanding some aspects of normal aging and Alzheimer's disease.

The theory we summarize here was formulated within the perspective of cognitive neuroscience. This approach to visual imagery led us to provide a description of information processing that rests on facts about the neural underpinnings of vision. We assume that complex cognitive abilities arise when different sets of component subsystems interact. Each such component carries out a single type of mapping between an input and output.

Components of Visual Imagery

It has long been known that visual mental imagery relies on some of the same mechanisms that are used in visual perception (for a good review of evidence, see Farah 1988). This is salubrious, given that we know a great deal about visual perception. Thus, we can gain a leg up on a theory of imagery by considering perception. In this section we provide a brief overview of the major features of the visual system, focusing on those aspects that involve stored information (so-called "high level" visual processes; for a more detailed treatment, see Kosslyn et al. 1990; Kosslyn 1991).

Visual Buffer

Introspectively, visual mental images appear to embody spatial properties of objects and scenes. For example, consider what seems to occur when you try to answer questions such as: "Which is longer, a donkey's ears or an ear of corn?" "Which is darker green, a Christmas tree or a frozen pea?" "Which is wider, a lightbulb or a tennis ball?" Most people report that they visualize the objects, and "see" the necessary properties. This sort of introspection suggests that visual mental images reconstruct the spatial properties of an object.

And in fact, numerous areas of cerebral cortex are spatially organized; patterns of activity within these areas make explicit the spatial organization of a stimulus. During visual perception, these areas represent images by mirroring the pattern of activation on the retina, and hence these areas are said to be "retinotopically organized." For example, Tootell et al. (1982) used a 2-deoxy-D-glucose (2DG) method to study the retinotopic organization of visual areas in Macaque monkeys. The monkeys were trained to stare at a flashing pattern and then were injected with a dose of 2DG. The more a given neuron worked when the monkey was staring at the pattern, the more of this radioactively tagged sugar it took up. After the animal stared at the pattern for 25–30 min, it was sacrificed and the pattern of radioactivity in its brain was examined. The radioactive trace of the sugar consumed by cortical nerve cells produced a pattern in area V1 (at the posterior end of the occipital lobe) that mirrored the actual pattern itself; it was as if there was a picture of the pattern laid across this area of cortex. Felleman and van Essen (1991) estimate that at least 15 cortical areas of the Macaque monkey have this type of spatial organization.

Fox et al. (1986) used positron emission tomography (PET) to show that area V1 in humans is also retinotopically organized. This is of interest because Kosslyn et al. (1991) showed that this area is active during visual mental imagery. Thus, we can infer that at least some of the spatial properties of visual mental imagery may arise when spatial patterns are activated from memory, not from the eyes, in spatially organized regions of cortex. We call these regions the "visual buffer."

Attention Window

There is far more information in the visual buffer than can be processed in detail. Thus, it is of interest that there appears to be an "attention window" within this buffer, which selects a region of the patterns of activity in the visual buffer and sends it to other areas for further processing (cf. Moran and Desimone 1985; Treisman and Gelade 1980).

Encoding Object Properties Versus Spatial Properties

The information within the attention window is further processed in two different pathways. One pathway goes from V1 ventrally to the temporal lobe, while the other pathway goes dorsally to the parietal lobe. To investigate what is processed in each pathway, Ungerleider and Mishkin (1982; see also Pohl 1973) trained Macaque monkeys to use either object properties or spatial information to find food. Different parts of the animal's brain were removed after the task was learned, and the behavioral deficits were observed. Removal of the temporal and parietal lobes resulted in a double disassociation: ablating the temporal lobes drastically affected an animal's ability to encode shape, but did not have much of an effect on its ability to encode location; in contrast, ablating the parietal lobes had the opposite effect. These results suggest that the dorsal pathway acts as a spatial encoding subsystem, whereas the ventral pathway acts as an object properties encoding components subsystem. Consistent with these inferences, neuroanatomical and neurophysiological studies have found neurons that are sensitive to shape and color in the temporal lobes (Desimone et al. 1984; Gross et al. 1984), and location and motion in the parietal lobes (Hyvarinen and Poranen 1974; Mountcastle et al. 1975; Anderson et al. 1985).

The division between encoding object properties (color and texture, in addition to shape) and spatial properties (size and orientation, in addition to location) makes sense from a computational point of view. Rueckl et al. (1989) trained a neural network computer model to distinguish between nine different shapes, each in nine different positions. A three layer feedforward network was used, which consisted of an input layer of 25 units (representing a 5×5 input matrix), a hidden layer of 18 units, and an output layer of 18 units; nine of the output units indicated the input shape (one unit per shape), and nine indicated the location (one unit per location). In one version of the model, all of the hidden units were connected to all of the output units; in another version, some of the hidden units were connected only to the output units that indicated shape, and the others were connected only to the output units that indicated location. Splitting processing into two distinct streams properly resulted in much better performance than having all functions take place in a single, undifferentiated system.

The split networks did especially well when more resources were allocated to the 'what' task (14 of 18 units); this finding suggests not only that the 'what' and 'where' functions are computationaly distinct, but also that identifying shapes is computationally more demanding than registering location – and in fact there are more neurons in the temporal object-properties system than in the parietal spatial-properties system (e.g., see van Essen 1985).

If visual memories reside in the object-properties system in the temporal lobe, as some have argued (for a review, see Kosslyn 1991), then visual images might be formed by activating these memories. This idea is plausible in part because the neural connections from lower-level visual areas to higher-level ones are reciprocal; they run in both directions. Thus, activating a visual memory in the temporal lobes could evoke a spatial pattern in the lower-level retinotopically organized areas. One reason that this would be necessary is that the higher-level visual areas in the temporal lobe are not spatially organized; thus, to reconstruct the spatial properties of objects, activity must be induced in lower-level, retinotopically mapped areas.

Associative Memory

The object- and spatial-properties encoding systems send information into the frontal lobe, which appears to have special short-term memories for different types of sensory input (Goldman-Rakic 1987). Furthermore, we can infer that there must be a long-term memory representation (probably not in the frontal lobes) that associates object properties with spatial properties; the mere fact that people can recall where furniture is located in their homes indicates that the two sorts of information must have been conjoined. This memory representation is multimodal, associating not only simply visual object properties and spatial properties, but also auditory, tactile, and other sorts of information.

This long-term memory representation is relevant to imagery in part because many complex objects are encoded over the course of numerous eye movements. In this case, parts may be encoded separately, and a representation of the structure of the object (indicating how parts are spatially organized) would be stored in associative memory. If so, then when such an object is visualized, this structure would be accessed to find the locations where the parts should be visualized.

Top-down Processing

During visual object identification, one sometimes cannot identify an object at first glance; the information encoded in a single eye fixation is not sufficient to match a stored representation. However, the input provides some indication of what is being viewed, and the partial match to stored representations may serve as a hypothesis. In this case, additional information is needed to make the identification. Frontal lobe processes apparently access information in associative memory and use this information to guide further eye fixations; area 8 of the frontal lobes (also known as the frontal eye fields) plays a major role in guiding eye movements, and many have found that damage to the frontal lobe disrupts systematic search (e.g., Luria 1980).

When visual mental images are formed, these processes access the stored representation of the structure of an object in associative memory and send information to the temporal lobes to activate images of the individual parts; the stored representation of spatial relations among the parts leads the system to form the image at the correct relative location for a part, so that the composite image is built up sequentially over time (for further details, see Kosslyn 1991). Kosslyn et al. (1988) provide evidence that images are in fact built up part-by-part over time.

Effects of Age on Imagery

We used our conception of imagery processing to guide investigations of the effects of increased age on visual mental imagery. Our approach to imagery suggests that a complex system forms and uses images, and some components are involved in some aspects of imagery but not others. Thus, we expected any factor that selectively affects components of the system to selectively affect different aspects of imagery.

Visual mental imagery, like other complex abilities, is not unitary and undifferentiated. Rather, it is useful to conceive of imagery as four distinct abilities. Consider the following task: you are loading luggage into the trunk of your car. You first study the pile on the curb, and then turn and visualize one suitcase in the trunk, and then another, and another. You imagine turning the suitcases, trying to fit them optimally into the trunk. While you are imaging placing one suitcase, you hold the images of the others where they seem best placed.

This task illustrates the four abilities, as follows. First, you had to *generate* an image of the suitcases; this involves activating visual memories. You not only activated the representations of shapes, causing a spatial pattern of activity in retinotopically mapped areas, but you also may have used stored descriptions in associative memory to help you arrange the shapes. For example, if you had packed those bags before, you might know that the big one nestles nicely against one side of a smaller one and fits nicely next to the spare tire; this information is used to direct where the images should be formed in the visual buffer.

Second, you had to *inspect* the image, scanning around and "seeing" whether the suitcases were properly placed; this involves shifting the attention window and encoding objects in the image – just as you would encode objects you see during perception. Thus, the object properties system is used both to form the image and then to match the subsequent input to visual memories, allowing you to recognize patterns that may have been only implicit in the image. For example, if asked if a cat has curved front claws, you might visualize the claws and then match the image to representations of curved vs. straight claws; never having thought about this before, you did not have an explicit representation of this property, but it was implicit in the image.

Third, you mentally *transform* objects in the image, moving them around; this function appears to involve parietal lobe and frontal lobe structures (Deutsch et al. 1988). Kosslyn (1987) suggests that the parietal lobes may be involved in shifting the location representation, but that this process is noisy; frontal lobe processes are involved in accessing information that is used to "clean up" the image. Thus, objects in images are transformed in small steps (see Shepard and

Cooper 1982), partly so that errors due to noisy shifting processes can be corrected before they are compounded.

Finally, as you scanned and transformed the objects in the image, you had to *maintain* the composite image. This process presumably involves repeatedly activating the visual memory in the temporal lobes, but does not require repeatedly accessing long-term associative memory. Maintaining an image can be achieved by re-encoding the image into the object properties system (as is done in visual perception), and then "bouncing" the pattern back into the visual buffer. Depending on how the imaged pattern is re-encoded, it will be more or less easily visualized; if one encodes it as relatively few perceptual units, less processing will be required than if one encodes it as relatively many perceptual units, each of which must be activated individually.

We examined the effects of age on these four abilities in young and old subjects (for a detailed report, see Dror and Kosslyn 1991). One group included 16 young subjects with an age range of 18–23 years (with a mean age of 20 years), and the other group included 16 elderly subjects with an age range of 55–70 years (with a mean age of 63 years). Each subject was tested on four tasks.

The first task tapped onto the image generation processes component. In this task the subjects were asked to generate an image of an uppercase letter within a set of four brackets placed at the vertices of an otherwise invisible rectangle. The subjects were to compare the imaged letter to a small "x" mark that was placed within the enclosed space, and to decide whether the letter would have covered the mark. We manipulated the difficulty of generating the images by varying the complexity of the letters. Kosslyn et al. (1988) found that images of more complex letters are more difficult to generate than simple ones. Within each level of complexity, half of the trials included 'x' marks that would have been covered by the letter, and half included 'x' marks that would not have been covered by it. We examined both the time to respond and the errors subjects made.

We found that both age groups had more difficulty imaging more complex letters in this task, replicating previous results, but this difference was larger for the older people. According to our theory, more complex patterns are imaged by (1) accessing a stored description of the arrangement of parts in associative memory, and then (2) forming images of the parts one at a time. Thus, the effects of age could reflect the process of looking up and using a description to arrange the parts or the process of activating the individual parts. Consistent with our theory, the subjects made more errors and required more time to evaluate probes that fell on segments that would be drawn late in the sequence of strokes than those that fell on segments near the beginning of the drawing sequence. In addition, the rate of increase in errors was much larger for older subjects than for young ones. Thus, we have evidence that the part generation process is affected by age.

The scanning task required the subjects to study a rectangular-ring shape, which resembled a square donut that was composed of black and white squares. After the subject studied the pattern of black and white squares, an arrow appeared within the ring for 50 ms. The arrow appeared in different places within the ring and was presented at eight orientations. Following the brief flash of the arrow, the display was removed. The subjects' task was to decide whether the arrow pointed to a black or to a white square. All of the rectangular-ring shapes included three black squares, which were located in different places along the perimeter. The arrow was positioned at three distances from the target square, which required the subjects to scan different distances; on half of the trials the arrow pointed to a black square, and in the other half it pointed to a white square.

The response times and errors increased with greater distances to be scanned, replicating previous results (see Kosslyn 1980). However, in contrast to the image generation findings, the increase in response time for greater distances was the same for both age groups, as was the increase in errors. Thus, we not only have evidence that the processes that control the location of the attention window during "image inspection" are distinct from those that generate images, but also that at least some image inspection processes are not affected by age in the same way as the image generation processes.

The image maintenance task was a variant of the generation task; instead of generating visual images from memory, this task required the subjects to retain an image. The subjects studied a pattern presented within a set of four brackets placed at the vertices of an otherwise invisible rectangle. The pattern consisted of one, two, or three perceptual units, formed by juxtaposing black squares to form bars and shapes. After the subject studied the pattern, it was removed; 2500 ms later an 'x' mark was presented within the four brackets, and the subjects were asked to decide whether the shape would cover the 'x' mark. On half of the trials, the shape would have covered the 'x' mark, and on half it would not have.

Both old and young subjects found this task more challenging when more complex stimuli were used, as indicated by increased response times and errors. However, the increase was the same when we compared the one-unit (a bar) and three-unit stimuli. The most interesting finding was for the two-unit stimuli. Young people evaluated them as quickly as one-unit stimuli, whereas old people evaluated them as slowly as the three-unit stimuli. One account for this finding is that young subjects could group two units and encode them as easily as a single bar, whereas old people could not. We assume that this grouping process reflects a "strategy" whereby top-down processes guide attention to group components. Thus, we conjecture that although top-down processes can shift attention as well in old age as in youth, they are less effective at other operations. This is intriguing because shifting attention probably relies on subcortical mechanisms (see Posner and Petersen 1990), whereas the grouping strategy may rely on cortical mechanisms.

Finally, the image rotation task required the subjects to decide whether two shapes were identical, regardless of their orientations. The two shapes were shown simultaneously, and had the same orientation (with no angular disparity) or an angular disparity of 90, 135, or 180 degrees. Half of the stimuli in the pairs presented at each orientation were identical, and half contained mirror-reversed versions of the stimuli. This task is a modified version of the task devised by Shepard and Metzler (1971), who found a linear increase in response time with increases in angular disparity.

Replicating previous results, we found that response times and errors increased with increased angular disparity, which presumably required the subjects to perform more "mental rotation" to align the figures prior to comparing them. In addition, the increase in errors with angular disparity was much more severe for the older subjects. Our finding of increased difficulties in mental rotation with age replicates that of Cerella et al. (1981) and Gaylord and Marsh (1975). Because we examined the slope of the increase, this finding reflects the processes that shift orientation per se, and not those involved in encoding the form.

In summary, the results suggest that older people have relative difficulty using top-down processing to access associative memory in the course of generating images, in using top-down processing to organize forms into relatively few perceptual units, and in rotating images. All of these processes appear to involve frontal lobe function. However, the older subjects could shift attention over an image as well as younger subjects, a process which we also speculate involves frontal lobe functions. There is at least one major difference between the role of the frontal lobes in scanning and in the other abilities: the other tasks all involve accessing stored information to guide a sequence of events. To generate an image, one must arrange a series of segments on an image. To maintain the image effectively, one must organize segments and then encode them. To rotate, one must shift an imaged form through a set of distinct increments. It is well known that the frontal lobe has a special role in sequencing events (e.g., Luria 1980), and we suspect that this function is affected by normal aging.

Alzheimer's Disease

Alzheimer's disease is characterized by a wide range of clinical disorders (Schwartz et al. 1990; Semple et al. 1982). Semple et al. (1982) described a number of distinct features of the disease, including memory deficit, spatial disorientation, aphasia, apraxia, and the deterioration of personality. If all of the clinical features of Alzheimer's disease are different manifestations of a single underlying cause, we are left with the puzzle of why such a cause would have such a wide range of effects. Our approach leads us to suggest that the disease can affect some parts of the brain more severely than others (cf. Reisberg 1983). Indeed, there may be different varieties of the disease, with different types affecting different neural structures most severely.

And in fact, Albert and Lafleche (1991) argue that there are two distinct subtypes of Alzheimer's disease; both are characterized by memory deficits, but one also has spatial deficits whereas the other also has language deficits. Presumably, these different deficit profiles arise because different parts of the brain have been affected. Indeed, it is possible that the spatial type of Alzheimer's disease represents a drastic extension of the normal process of cognitive aging. If so, then structures that suffer relatively early during the course of normal aging may be particularly susceptible to the effects of the disease.

Our findings suggest that at least some subcortical structures may be relatively resistant to the effects of aging. Posner and Petersen (1990) argue that the superior colliculus is involved in shifting attention, and we found no deficit in image scanning over normal aging. This finding may suggest that the subcortical structures responsible for scanning are more resistant to the effects of neural deterioration than are certain cortical areas, which may also imply that the abilities these structures support would be relatively intact in the spatial type of Alzheimer's disease. To our knowledge, this possibility has not yet been investigated.

On the other hand, we have hypothesized that certain frontal lobe functions may be impaired over the course of normal aging. If the spatial type of Alzheimer's disease is a severe extension of the normal aging processes, then we would expect similar deficits in these patients. Although Alzheimer's disease is not usually characterized in terms of its effects on frontal lobe function (Martin 1990), Alzheimer's patients often do in fact show signs of frontal lobe damage late in the course of the disease (Schwartz and Chawluk 1990). Indeed, at least part of their difficulties in naming objects may reflect difficulties in the process of accessing information from long-term memory, which may depend on frontal structures (cf. Luria 1980). Moreover, some Alzheimer's disease patients show deficits associated with frontal lobe functions early in the course of the disease (Albert and Lafleche, 1991). These deficits include problems in sequencing, monitoring, and shifting behavior.

In addition, Alzheimer's disease apparently can selectively affect parietal and temporal cortex, which reportedly has distinct effects on spatial orientation and language (Appell et al. 1982; Cummings et al. 1985). It is of interest to discover whether normal aging affects these structures to the same *relative* degree as Alzheimer's disease.

In short, it would be of great interest to compare spatial deficits in normal aging and Alzheimer's disease, using fine-grained tasks that assess distinct aspects of spatial processing. It is possible that the spatial variety of Alzheimer's disease simply amplifies normal trends; alternatively, it may disrupt processing in ways that do not occur in the course of normal aging.

Conclusions

Research on visual mental imagery suggests that a set of component subsystems works in concert to produce various aspects of imagery; different combinations of processes are used to generate, maintain, inspect, and transform images. We have offered hypotheses about which brain structures support each component process, and summarized some initial evidence that normal aging affects some components more than others. It will be of interest to discover whether measures of cortical and subcortical degradation in different varieties of Alzheimer's disease allow us to predict particular profiles of imagery deficits.

In all likelihood, the present theory will be useful for understanding some aspects of these patient's deficits, but fall short in providing a detailed understanding. Thus, trying to use the theory to understand their deficits might not only illuminate aspects of Alzheimer's disease, but also might enrich the development of the theory more generally.

References

Albert MS, Lafleche G (1991) Neuropsychological testing of memory disorders. Massachusetts General Hospital, Boston
Anderson RA, Essick GK, Siegel RM (1985) Encoding of spatial location by posterior parietal neurons. Science 230: 456–458
Appell J, Kertesz A, Fisman M (1982) A study of language functioning in Alzheimer patients. Brain Lang 17: 73–91
Cerella J, Poon LW, Fozard JL (1981) Mental rotation and age reconsidered. J Gerontol 36: 620–624
Cummings J, Benson D, Hill M, Read S (1985) Aphasia in dementia of the Alzheimer type. Neurology (NY) 35: 394–396
Desimone R, Albright TD, Gross CG, Bruce CJ (1984) Stimulus selective properties in inferior temporal neurons in the macaque. J Neurosci 4: 2051–2062
Deutsch G, Bourbon WT, Papanicolaou AC, Eisenberg HM (1988) Visuospatial tasks compared via activation of regional cerebral blood flow. Neuropsychologia 26: 445–452
Dror IE, Kosslyn SM (1991) Imagery and aging. Harvard University, Cambridge
Farah MJ (1988) Is visual imagery really visual? Overlooked evidence from neuropsychology. Psychol Rev 95: 307–317
Felleman DJ, van Essen DC (1991) Distributed hierarchical processing in primate cerebral cortex. Cereb Cortex (in press)
Fox PT, Mintun MA, Raichle ME, Meizen FM, Allman JM, van Essen DC (1986) Mapping human visual cortex with positron emission tomography. Nature 323: 806–809
Gaylord SA, Marsh GR (1975) Age differences in the speed of a spatial cognitive process. J Gerontol 20: 674–678
Goldman-Rakic PS (1987) Circuitry of primate prefrontal cortex and regulation of behavior by representational knowledge. In: Plum F, Mountcastle V (eds) Handbook of physiology, vol 5. American Physiology Society, Washington, pp 373–417
Gross CG, Desimone R, Albright TD, Schwartz EL (1984) Inferior cortex as a visual integration area. In: Reinoso-Suarez F, Ajmone-Marsan C (eds) Cortical integration. Raven, New York
Hart RP, Kwentus JA, Harkins SW, Taylor JR (1988) Rate of forgetting in mild Alzheimer's type dementia. Brain Cogn 7: 31–38
Hyvarinen J, Poranen A (1974) Function of the parietal associative area 7 as revealed from cellular discharge in alert monkeys. Brain 97: 673–692
Kosslyn SM (1980) Image and mind. Harvard University Press, Cambridge
Kosslyn SM (1987) Seeing and imaging in the cerebral hemispheres: a computational approach. Psychol Rev 94: 148–175
Kosslyn SM (1991) A cognitive neuroscience of visual mental imagery: further developments. In: Logie R, Denis M (eds) Advances in imagery research. Erlbaum, London
Kosslyn SM, Cave CB, Provost D, von Gierke S (1988) Sequential processes in image generation. Cogn Psychol 20: 319–343
Kosslyn SM, Flynn RA, Amsterdam JB, Wang G (1990) Components of high-level vision: a cognitive neuroscience analysis and accounts of neurological syndromes. Cognition 32: 203–278
Kosslyn SM, Alpert NA, Maljkovic V, Weiss SJ, Hamilton SE (1991) Visual mental imagery affects processing in primary visual cortex: a PET study. Harvard University, Cambridge
Luria AR (1980) Higher cortical functions in man. Basic, New York
Martin A (1990) Neuropsychology of Alzheimer's disease: The case for subgroups. In: Schwartz MF (ed) Modular deficits in Alzheimer-type dementia. MIT Press, Cambridge
Moran J, Desimone R (1985) Selective attention gates visual processing in the extrastriate cortex. Science 229: 782–784

Mountcastle VB, Lynch JC, Georgopoulos A, Sakata H, Acuna C (1975) Posterior parietal association cortex in the monkey: command functions for operations within extrapersonal space. J Neurophysiol 38: 871–908

Pohl W (1973) Dissociation of spatial discrimination deficits following frontal and parietal lesions in monkeys. J Comp Physiol Psychol 82: 227–239

Posner MI, Petersen SE (1990) The attention system of the human brain. Annu Rev Neurosci 13: 25–42

Reisberg B (1983) An overview of current concepts of Alzheimer's disease, senile dementia, and age associated cognitive decline. In: Reisberg B (ed) Alzheimer's disease: the standard reference. Free Press, New York

Rueckl JG, Cave KR, Kosslyn SM (1989) Why are 'what' and 'where' processed by separate cortical visual systems? A computational investigation. J Cogn Neurosci 2: 171–186

Schwartz MF, Chawluk JB (1990) Deterioration of language in progressive aphasia: A case study. In: Schwartz MF (ed) Modular deficits in Alzheimer-type dementia. MIT Press, Cambridge

Schwartz MF, Baron B, Moscovitch M (1990) Symptomatology of Alzheimer-type dementia: report of a survey by mail. In: Schwartz MF (ed) Modular deficits in Alzheimer-type dementia. MIT Press, Cambridge

Semple SA, Smith CM, Swash M (1982) The Alzheimer disease syndrome. In: Corkin S, Davis KL, Growdon JH, Usdin E, Wurtman RJ (eds) Alzheimer's disease: a report of progress in research. Raven, New York

Shepard RN, Cooper LA (1982) Mental images and their transformations. MIT Press, Cambridge

Shepard RN, Metzler J (1971) Mental rotation of three-dimensional objects. Science 171: 701–703

Tootell R, Silverman M, Switkes E, DeValois R (1982) Deoxyglucose analysis of retinotopic organization in primate striate cortex. Science 218: 902–904

Treisman AM, Gelade G (1980) A feature integration theory of attention. Cogn Psychol 12: 97–136

Ungerleider LG, Mishkin M (1982) Two cortical visual systems. In: Dingle DJ, Goodale MA, Mansfield RJW (eds) The analysis of visual behavior. MIT Press, Cambridge

Van Essen DC (1985) Functional organization of primate and visual cortex. In: Peters AA, Jones EG (eds) Cerebral cortex. Plenum, New York

Neuronal Models of Cognitive Functions Associated with the Prefrontal Cortex

J.-P. Pierre Changeux and *S. Dehaene*

Summary

Understanding the neural bases of cognition has become a scientifically tractable problem, and neurally plausible models are proposed to establish a causal link between biological structure and cognitive function. To this end, levels of organization have to be defined within the functional architecture of neuronal systems. Transitions from any one of these interacting levels to the next are viewed in an evolutionary perspective. They are assumed to involve:
1. the production of multiple transient variations and
2. the selection of some of them by higher levels via the interaction with the outside world.

The time scale of these "evolutions" is expected to differ from one level to the other. In the course of development and in the adult, this internal evolution is epigenetic and does not require alteration of the structure of the genome. A selective stabilization (and elimination) of synaptic connections by spontaneous and/or evoked activity in developing neuronal networks is postulated to contribute to the shaping of the adult connectivity within an envelope of genetically encoded forms. At a higher level, models of mental representations, as states of activity of defined populations of neurons, are suggested and their storage is viewed as a process of selection among variable and transient "pre-representations". Models are presented which can perform the delayed response task or the Wisconsin card sorting test and cognitive functions such as short-term memory, reasoning and handling of temporal sequences. Implementations of these mechanisms at the cellular and molecular levels are proposed. Finally, speculations are offered about plausible neuronal models and selectionist implementations of intentions.

Introduction

In biology, as in physics, the theoretical approach precedes experiment. Knowledge progresses by "conjecture and refutation" (Popper 1963), by the construction of *models* followed by submitting them to the experimental test. Any model is a "representation" of a natural object or process described in a coherent, non-contradictory and minimal form, if possible mathematical. To be useful, the way in which it is formulated must allow comparison with outside reality. However, it cannot be expected that it will offer an exhaustive description of the latter. It may eventually be adequate, but will always remain limited.

Claude Bernard introduced a major distinction into life sciences by comparing anatomy (stable morphological organizations or "structures") with physiology (the dynamic processes by which an organism acts on the outside world or on itself). From there, the purpose of life sciences has become more precisely the determination of causal relationships between structure and function. This determination of relationships acquires a new dimension in the case of higher brain functions, where what is revealed of the mental state is still very often deliverately dissociated from subjacent neural organizations. Our purpose is completely opposed to any approach of this type since, in contrast, it concerns the creation of "a bridge" between neural sciences and mental sciences by developing neuronal models of cognitive functions.

Under these conditions, models of this type must be plausible at the neuro-biological level and not merely be "artificial", which would exclude from the start any comparison with neuronal reality. To be adequate, it is also necessary that the strucure-function relationship should provide correspondence between theoretical and experimentally observable variables in a *pertinent* manner and particularly *at the level of organization* involved (Changeux and Dehaene 1989).

As the physician P. Anderson wrote in 1972 "the ability to reduce everything to simple fundamental laws does not imply an ability to start from these laws and reconstruct the universe." The organization of living beings must be considered within the context of the evolution of species, and this evolution shows a consistent increase in complexity during palaeontological history. A first theoretical and essentially qualitative approach is to break down these complex organizations into hierarchical levels of organization whereby their appearance during evolution coincides with the appearance of new functions. For F. Jacob (1970), "living beings are therefore constructed by means of a series of packages. They are fitted together according to a hierarchy of discontinuous units or "integrons." Nerve cells are composed of molecules, but are assembled together into networks by dendrites and axons. The extension of this paradigm to clusters of neurons leads to differentiation of levels of organization within the brain itself. This must not be confused with the "levels" which Marr (1982) distinguishes in "Vision":

1. the "hardware" or neural machine;
2. the representation and algorithm and,
3. the computational theory.

In this case it concerns *levels of understanding* according to a scheme which perpetuates the cleavage between structure and function. This distinction may be useful for defining experimental pathways of approach or even for describing the system and its functional properties. However, it by no means takes any account of the levels *of organization* or of integration properly speaking and which may each highlight, at least in part, a description according to the terms of Marr.

Several models of the cleavage of functional brain organization into distinct "levels of integration" may therefore be proposed. Initially, the following levels were separated:

1. the level of elementary circuits and simple reflexes or fixed schemes of action (Kant's "sensitivity"?);
2. that of "groups of neurons" and of "symbolic representations" (Kant's "intendment"?);
3. that of complex assemblies of neuronal groups that we may refer to as "reason" (Kant) a "knowledge" (Newell) level (Newell 1982; Dehaene and Changeux 1989). However, we must expect even finer hierarchical cleavages.

The transition from one level of organization to the next is considered within the general conceptual context, which in this laboratory (Changeux et al. 1973; Changeux 1983) has always been that of an evolutionary epistemology (Popper 1966; Campbell 1974) inspired from Darwin's ideas (Darwin 1859; Poincare 1913) based upon:

1. a "blind" "generator of diversity" which introduces "variations" into the functional organization at the level being considered;
2. a mechanism of conservation and/or of propagation of the selected variation.

Within the context of the Darwinian scheme of evolution of species, variation occurs at the genome level (mutations, chromosome rearrangements, etc.), and the conservation of variations is performed by conservative replication of DNA and by propagation via sexual reproduction.

The application of this paradigm to the "internal" levels of organization of the brain does not postulate covalent variation of the genome but, in contrast, "epigenetic" variations of connectivity during development (time scale: years, minutes) or states of activity of neuronal clusters at levels of symbolic representation or architectures of reason (time scale: 0.1 second, minute). The proposed models (Dehaene and Changeux 1989, 1991) relate to the general problem of transition from the "symbolic" level to the "reason" level (Changeux and Dehaene 1989), dealt with in the particular example of the prefrontal cortex.

Functional Organization of the Prefrontal Cortex

The prefrontal cortex is the region of the neocortex in which the surface area has increased relatively most during the course of mammalian evolution; it increases from 3.5 % in the cat to 17 % in the chimpanzee and, finally, 29 % in humans (see Fuster 1989). Its histological organization is not uniform. In the 19th century, Brodmann subdivided the prefrontal cortex into several distinct territories, essentially on an anatomical basis. However, the latter mainly shows the six layers characteristic for the entire associative cortex, with a well-differentiated granular layer IV. The connectivity of the hundreds of millions of neurons which compose it (900 millions according to some estimates) is one of extreme richness; this is most often established in a reciprocal or "re-entrant" manner according to Edelman's (1978) term between the intrinsic neurons which compose it, but also with the neurons of many regions of the encephalon. The frontal cortex exchanges connections with regions of the cortex which are hierarchically inferior to it, parieto-temporal associative areas, then primary sensory areas and motor areas. Of all the cortex, it is the domain which is most densely connected with the limbic system, which, as is known, participates in emotional responses. It is also linked to the thalamus as well as to the basal ganglia, which are involved in the control of movement. Finally, several nuclei of the reticular formation (containing neurotransmitters such as dopamine, noradrenaline, etc.) send highly divergent axons towards the frontal cortex where they control activity in a "global" manner. Examination of the connectivity of the frontal cortex allows at least three levels of hierarchical organizations to be defined, "nested" (Campbell 1974) one within the other. Very schematically, the prefrontal cortex is surrounded by an inferior level which corresponds to the areas of association, and by a more global level represented by the reticulo-frontal loops.

As Harlow noted in 1868, lesions of the frontal lobe in humans are accompanied by emotional disorders (hyperemotivity, character instability; Nauta 1971) as well as profound "cognitive" disorders which are expressed both by excessive obstinacy (perseverance in error) and then abnormal tendency to distraction, with a general decrease in critical activity. For Diamond (1988) the frontal cortex ensures "the correlation of information with space or time" and "inhibits dominant action tendencies." It constructs and updates "representations of the environment" (Goldman-Rakic 1987) which allow the subject to "plan and elaborate anticipations" (Teuber 1964, 1972) of actions on the surrounding world. For Shallice (1982) it constitutes a system of "supervisory attentive system", hierarchically superior to the "routine" "contention scheduling" system. It ensures the construction of plans in non-routine situations and *selects* schemes appropriate to those situations, all the while recording and taking into account errors likely to intervene in the realization of the plan. The prefrontal cortex produces "mental syntheses" (Bianchi) and is the site of "intentional behaviour" (Pavlov). It is also involved in making decisions and forming plans in relation to social behavior (Damasio et al. 1990).

It may be suggested that it corresponds to the "knowledge level" which theoreticians of artificial intelligence (Newell 1982) locate *above* the "symbolic" level and which may be the homologue of the "reason" level. Under these conditions, the prefrontal cortex would participate in the neural *architectures of reason* (Changeux 1988).

Functional Analysis of the Prefrontal Cortex by Various Tasks with Delayed Responses

In 1914, Hunter developed a behavioral test called the "delayed response task" which has since been very widely used with laboratory animals and even with children for the experimental analysis of prefrontal functions (Diamond 1988; Piaget 1954; Fuster 1984; for review see Dehaene and Changeux 1989). The experimental design is as follows. A stimulus or cue object is initially presented to the subject at a precise point in a scene, then a screen falls and covers the scene from the subject's sight for a variable duration; two objects are then presented at two different sites and the subject must choose one of the two. The rule defining the correct choice varies with the type of task involved. In the delayed response (DR) task in its strict sense, and in task A$\bar{B}$ (A not-B; Piaget 1954), the rule is to choose the object which is located in the *position* occupied by the cue before the interval. In the DR task, the position of the cue is changed at random from one test to another, whereas in task A$\bar{B}$ its position is changed only after the child has been successful in accomplishing the task. In the so-called "delayed matching-to sample" task (DMS), the subject must choose an object identical to the cue irrespective of its position. Finally, a third task, called delayed alternation (DA), may be considered as part of this group of tasks. After having successfully performed a task at a given position, the subject must choose the alternative position in the following response. In all of these protocols, the subject learns the task during a training phase in which he receives the reward for each successful test (fruit juice in the monkey, playing with a toy in the case of children, etc.).

In all cases these are sensory-motor tasks which involve short-term memory of the subject and require selective attention. During the task, the subject makes a *decision* by comparing a test object with a memorized representation of the cue object. Finally, during training the subject performs an *induction* process in time and space by discovering the abstract rule (pertinent choice of feature) which governs the reinforcement.

In young children, systematic success in the A$\bar{B}$ test develops around the age of 7.5 months and performances improve up to 12 months. The young macaque monkey masters these two tests between 1.5 and 4 months, and ablation of the frontal cortex at birth interferes with mastery of the test. In the absence of the delay, an immature subject or injured adult passes the test but, if the delay exceeds 1–2 seconds, performance deteriorates and essentially becomes random (for review see Diamond 1988).

The delayed response task reveals early "cognitive" functions which are nevertheless of a high order and linked to the integrity of the prefrontal cortex.

Electrophysiological Recordings During the Execution of a Delayed Response Task

To our knowledge, only a few rare electrophysiological data exist on the *acquisition* of mastery of the delayed response task (Kubota and Komatsu 1985). The principal data available are recordings of individual neurons in the monkey (macaque) *during the performance* of the DR test after training (for review see Fuster 1989; Watanabe 1986a). Neurons of a first type come into action when the cue is presented (Fuster 1973; Niki 1974; 1975). Their activity represents either an invariant early response appropriate to the *task*, which relates to the focussing of attention on the cue, or a response to the *test* itself, which distinguishes one cue from another, or finally, in some cases, both. Neurons of a second type, most often excitatory, change their activity in relation to the execution of the task; in general, their activity varies with the direction of movement of the hand which is required in the execution of the task (Kubota and Niki 1971; Watanabe 1986b). Their most remarkable feature is that their activity may anticipate the motor response by several seconds. Finally, the neurons of a third type are permanently active during the delay period, sometimes for a minute or more. Cells of this type are also encountered in the medio-dorsal thalamus and also in the temporal cortex, but to a much more limited degree. Neurons of this type are only observed in animals which have undergone training (it therefore does not concern a delayed sensory discharge). In addition, their activity is related to the state of *alertness* of the animal. Finally, there is a correlation between the activity of the cells during the delay period and the success of performance (distraction of the animal by an auditory stimulus during the delay period interferes both with the delay period activity and with success in the test). These neurons therefore establish a "temporal contingency" between presentation of the cue and motor performance. Their activity is not uniform. The discharge of a fraction of these cells tends to diminish during the delay period; these are *short-term memory cells* which participate in retention of the "representation" of the stimulus. The frequency of discharge of another fraction of neurons increases with time; they correspond to *motor anticipations* which prepare for movement. Coordination of the activity of these two categories of neurons ensures integration between short-term memory and preparation for motor action. Some cells which become temporarily active during the delay periods may participate in this coordination (Batner et al. 1981). At the end of the test, some prefrontal cells become active when the animal receives the reward and drinks the few drops of fruit juice received for a correct response. In contrast, other cells become active when the reward is expected but *is not* received (Rosenkilde et al. 1981).

These different types of cells are found throughout the prefrontal cortex; some are present in greater abundance in certain areas. The "reward" cells are present in the orbital regions, richly connected to the limbic system. The cells which are activated (or inhibited) during the delay period are mainly found in the region of the main sulcus of the dorso-lateral prefrontal cortex (Fuster 1989).

A Model of the Formal Neuron Network Accomplishing the Delayed Response Test

The objective of this modelling (Dehaene and Changeux 1989) is to construct a minimal and biologically plausible neuronal network which successfully passes the delayed-response task. Such a network must allow identification of the critical elements which are *necessary* for success in the task, and prediction of new properties subject to experimental validation.

The Formal Organism and Its Environment

The formal neuronal network is contained in a "formal organism" which interacts with a very restricted environment. This environment is *a priori* limited initially to the objects serving as a cue, then to some pertinent features of the latter which are likely to be taken into account by the formal organism: position (dimension 1), with two possibilities, right or left; color (dimension 2) with three possible hues; and finally no more than two objects may be presented to the formal organism at any given moment.

Each task is composed of successive tests and each test comprises four stages: presentation of an object, interval in the absence of the cue object, presentation of two objects, choice of object with reward or punishment, and an interval between two tests.

In tests of type 1 (analogues of DR or A$\bar{\text{B}}$), the correct object is the one having a position (dimension 1) coinciding with that of the cue; in those of type 2 (analogues of DMS), the correct choice is that of color of the cue (dimension 2).

The reward (or punishment) signal is applied from outside (by a master who decides his mark) or, under more natural conditions, as a result of the sensory qualities (taste, nutritional value, etc.) which are intrinsic to the object and recognized by the organism as favorable (or unfavorable) to survival (as a result of its past evolution). The reinforcement parameter covers an interval $[-1, +1]$ where 0 is neutral, +1 is maximum reward and −1 is maximum punishment.

Elementary Components of the Network

The network is composed of *"formal neurons"* of the McCulloch and Pitts type (1943) (see Amit 1989 for discussion), linked together by synaptic contacts of either the excitation or the inhibitory type. Each neuron is able to exist in two states, active (discharge) or inactive (rest). However, the states of activity of individual neurons (or synapses) are not explicitely modelled.

The basic unit of the network is a *"cluster of synergic neurons"*, the state of activity of which is assumed to code for an elementary "neural representation." This latter is analogous to Mountcastle's elementary module or "column" (Mountcastle 1978) or to Edelman's "group" of neurons (1978). It is defined and formalized here (Dehaene et al. 1987) as one hundred (or several hundred) neurons densely interconnected by excitatory synapses and, because of this, likely to exist in two self-sustained states of activity with either a high or a low frequency of discharge.

The neuron clusters are linked together by *"axon bundles"* of two types. The static bundles, not modulated by the activity of the network, propagate either lateral inhibition between clusters or the output of calculations performed by groups of clusters (or assemblies). The efficacy of the *modulated bundles*, for example between A and B, is regulated (to a *maximum* value) by the activity of a third neuronal cluster, for example C, called a modulator. The maximum efficacy value reached is itself variable and regulated by training (see below).

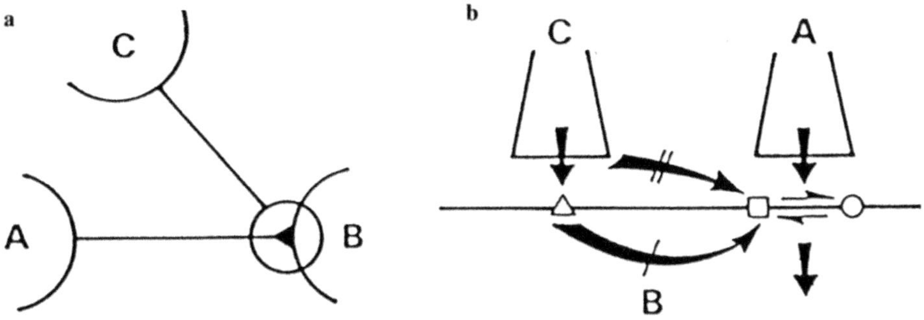

Fig. 1. Synaptic triad. Signals from synapse *C-B* modulate the efficacy of the neighboring *A-B* synapse. *b* is an enlarged view of *a* (taken from Dehaene et al. 1987)

Regulation of synaptic efficacy between A and B is undertaken in a "hetero-synaptic" manner by C according to the *"synaptic triad"* scheme (Fig. 1); Dehaene et al. 1987) where the signal produced by synaptic terminal C acting on neuron B regulates, by an extra- or intracellular signal, the allosteric transitions (Heidmann and Changeux 1982; Changeux and Heidmann 1987) of the postsynaptic receptor of synapse A → B. All the synaptic triads between neurons belonging to clusters of neurons A, B or C constitute a "modulated bundle."

Architecture of the Network

A major feature of the architecture of the network is its differentiation into two hierarchical levels of organization (Fig. 2). *Level 1* (or execution level) includes two layers of neuron clusters, respectively *"input"* and *"output."* Each of the characteristic features of a given object is analyzed and coded by a particular cluster of input neurons. The output clusters are connected in an isomorphic manner with the input clusters, and the activity of the output clusters governs the orientation of the organism towards a defined object possessing a particular feature. *Level 2* (or regulation level) includes a layer of "memory clusters" and a layer of "rule-coding clusters," and controls the processing of an object according to a defined rule. The memory clusters, which are self-excitatory and mutually inhibitory, project isomorphically and modulate the input-output connections. Each cluster of rule-coding neurons codes not for a particular feature of the object but for *one* dimension which groups together several features of the object (in the very restricted case of the proposed model, there are only two). The clusters of rule-coding neurons project onto bundles which link input clusters with memory clusters and regulate their efficacy. By analogy with the primate neocortex, level 1 would correspond to a visuo-motor loop which includes secondary visual areas and the motor or pre-motor cortex, and level 2 would be identified with the prefrontal cortex.

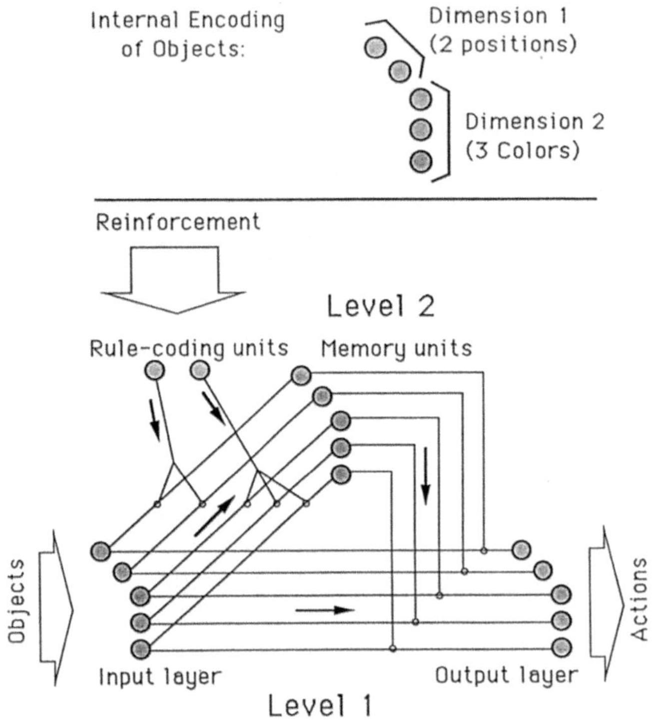

Fig. 2. Model of the role of the frontal cortex in learning and execution of delayed response tasks (taken from Dehaene and Changeux 1989)

Learning a Behavioral Rule

The organism learns a defined behavioral rule by interpreting a reinforcement signal which governs both modifications of synaptic efficacy and random variations in spontaneous activity of clusters of rule-coding neurons.

The reinforcement triggered "by return" as a result of the action of the organism on the environment during the learning process is "internalized" in the form of a parameter R which represents satisfaction (from 0 to +1) or dissatisfaction (from 0 to −1) of the organism. A first effect of R is to modulate the *maximal efficacy* of a synaptic triad according to Hebb's law. When R is positive and the postsynaptic neuron B (Fig. 1) is active at the same time, maximal efficacy increases; it decreases when the postsynaptic neuron is inactive. When R is negative, the rule is reversed.

Application of this rule is based on the allosteric properties of the postsynaptic receptor of synapse A → B. It is known that the nicotinic receptor for acetylcholine can exist in at least two desensitized states for which the ionic channel is closed (Changeux and Heidmann 1987). State I, of rapid access from the resting state, is involved in the functioning of the synaptic triad. The fraction of receptors in state D, of slower access, determines the maximum amplitude of variation in synaptic efficacy and is stabilized by the co-occurrence of two signals:

1. a *postsynaptic* signal (for example the intracellular concentration of Ca^{++}) which indicates recent activation of the cell, and
2. a diffuse *extracellular* signal (for example, the catecholamines of divergent reticulo-frontal pathways) which is propagated throughout all the synapses of the network for instance by "volume transmission" in the extracellular spaces (Fuxe and Agnati 1991; see Dehaene and Changeux 1991).

A second effect of R is to modify the activity of clusters of rule-coding neurons. When the organism is dissatisfied, R becomes negative, there is destabilization of all the rule-coding clusters, and spontaneous activity then varies from one cluster to another.

Learning takes place by selecting a defined cluster of rule-coding neurons according to its actual state of activity. The layer of rule-coding clusters serves, in some way, as a Darwinian "generator of diversity" and its evolution in time is under the control of the reinforcement signal.

Functional Properties of the Model

Simulation of the behavior of a network comprising only level 1 shows that the organism which possesses it is able to learn as a result of the action of the reinforcement loop on the triads of the input-output "execution" network. In an A$\bar{B}$ test, it ceases to orient at random. There is acquisition of a systematic orientation towards position A for which it was trained when A and B were presented simultaneously. However, like all infants or monkeys before the development of frontal connections, there will be systematic error when the position of the cue is changed from A to B. The organism which possesses only level 1 thus fails in the DR and DMS tasks. In contrast, it succeeds in all these tasks when it possesses levels 1 and 2.

The rule-coding neurons play a decisive role in the behavior of the organism. Their activity commands the memorization of a particular feature of the cue by modulating the efficacy of the connections between input clusters and memory clusters. If the rule-coding neurons which code for color are active, only the particular color of the cue will be memorized, but not its position. The neurons of the memory group themselves will govern the orientation of the organism towards the object possessing the memorized feature. In other words, the organism selects the object which possesses the characteristic feature of the cue *to the extent* that the rule-coding neurons which code for the particular *category* (position, color) to which this feature belongs are active.

The activity of the rule-coding neurons definitively "channels" the rule of behavior of the organism towards the choice. Learning therefore consists of a *search* among the various states of activity of rule-coding clusters to find the particular state which leads to the satisfaction of the organism. During learning, by successive "anticipations" based on the spontaneous variable and "blind" activity of rule-coding neuron clusters, the organism tests various features of the environment and selects the particular category of features of the object for which the "reward" is systematically positive.

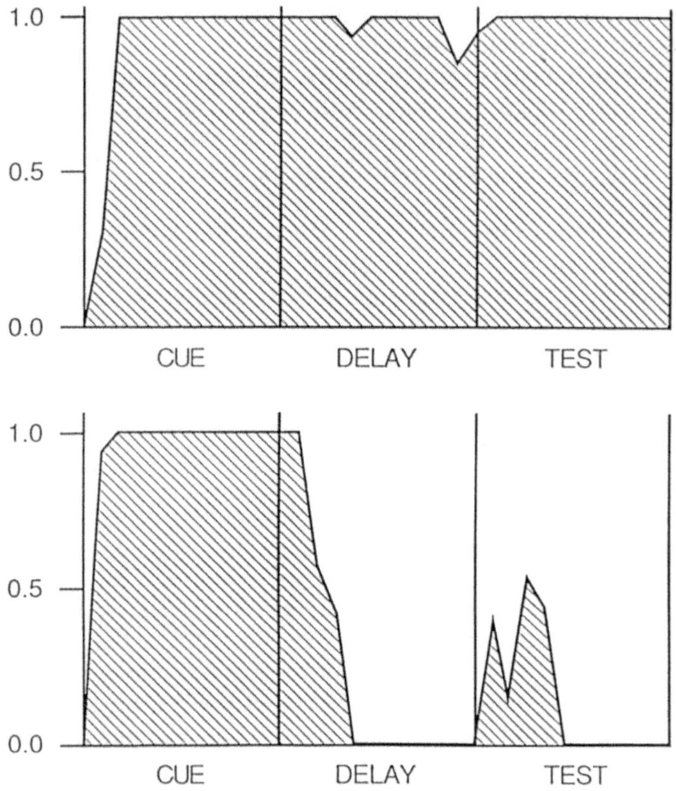

Fig. 3. Simulation of the activity of a memory neurons cluster during the delay. *Top*, the group remains active during the delay; performance in the next test is correct. *Bottom*, the group is inactivated due to internal noise; the organism now fails the test (taken from Dehaene and Changeux 1989)

The model allows simulation of the activity of defined neuron clusters during or after learning. In particular, neurons of the memory clusters display an activity which resembles that of neurons which are active during the delay period (see previous chapter) and the activity of which anticipates the behavior of the monkey during the choice when it is successful (but also when it fails; Fig. 3).

Simulation of the behavior of the formal organism shows that, with level 1 only, its behavior is analogous to the performances of infants aged from 7.5 to 9 months, of monkeys of 1.5 to 2.5 months or of monkeys with prefrontal lesions 17 (Fig. 4).

With level 2, the performances of the formal organism become practically identical to those of a child aged 12 months or of a rhesus monkey aged 4 months with respect to the learning of task A$\bar{B}$ or DMS. In addition, the organism is capable of passing from one task to another without difficulty.

Despite these successes, the formal organism modelled in this way presents three groups of limitations:

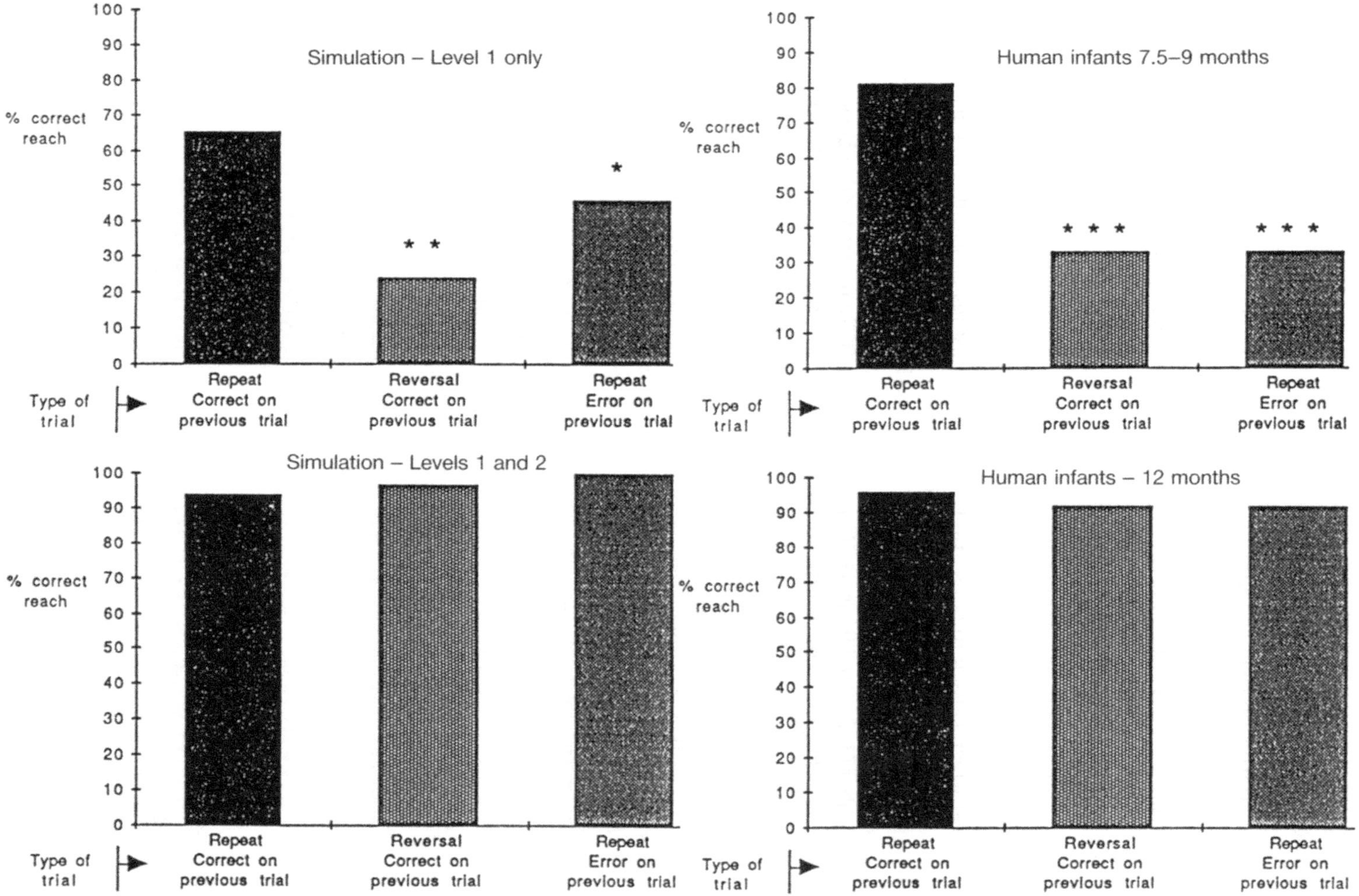

Fig. 4. Comparison of the performance of the network with that of children tested by Diamond (1985). The network with level 1 only is comparable to children before the development of frontal connections. The network with levels 1 and 2 succeeds just as well as older children (taken from Dehaene and Changeux 1989)

1. the sensory-motor tasks are highly reduced in the number of sensory categories or features and types of motor behavior;
2. the architecture is extremely simple: the number of formal neurons is six-seven orders of magnitude lower than that of the neurons of the prefrontal cortex in humans; and
3. the range of available rules is very small in size.

With the purpose of extending this research to more complex functions and to richer networks, a modeling of the Wisconsin card sorting test was investigated (Dehaene and Changeux 1991).

The Wisconsin Card Sorting Test

This test which is used to detect prefrontal cortex lesions classically consists of discovering the principle according to which a deck of cards must be sorted (Grant and Berg 1948; Milner 1963). The cards bear geometric figures of different shapes (triangle, star, cross or circle), color (green, red, blue or yellow) and number (1, 2, 3 or 4 figures). Four *reference* cards are permanently placed in front of the subject. The subject has another deck of cards called the *response* cards. He is asked to match each response card successively with one of the four reference cards. After each response, he is told whether this is correct or not. The subject tries to achieve the maximum of correct responses. The rule will, for example, be sorting according to color. Once the subject is systematically successful in this, the rule is changed, for example from color to shape. The subject must understand that the rule has changed and discover the new rule (Fig. 5).

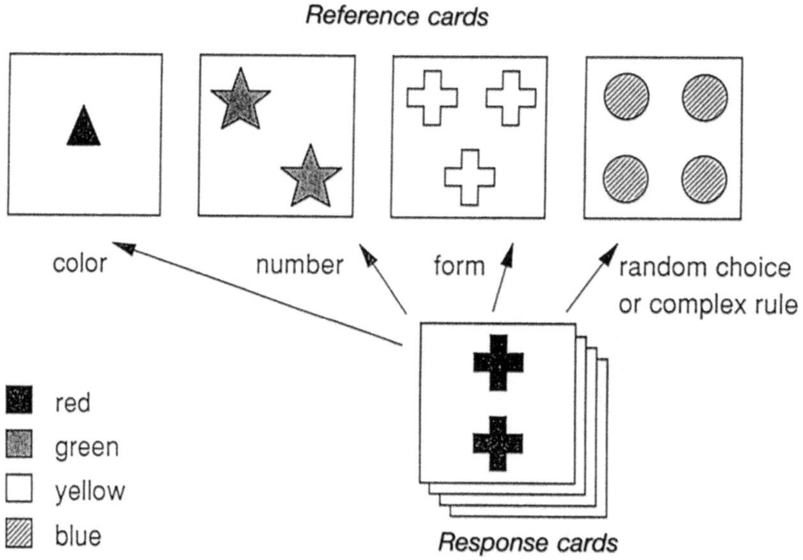

Fig. 5. Cues used in the Wisconsin card sorting test (taken from Dehaene and Changeux 1991)

Normal subjects do not always succeed in passing the test, particularly in the case of elderly subjects. However, subjects with a prefrontal lesion are systematically less successful than normal subjects. Frontal subjects make errors of a particular type called "perseveration" when they persist in using a rule which was initially correct even after they have been told that it is no longer in use. They exhibit difficulties in passing from one rule to another. The lesions which lead to the most marked deficit are located in the median frontal cortex.

Funtional analysis of the abilities of a formal cognitive system to pass the test (Dehaene and Changeux 1991) leads to the distinction of 6 "formal machines" according to the manner in which they select a new rule (Fig. 6).
The first three machines are "blind" in the sense that, each new rule is drawn at random from a repertoire of available rules without resort to reasoning. The simplest possible machine (random) draws a new rule entirely at random to replace the previous rule. The second, more complex [random + context] machine avoids again drawing a rule which has just been rejected. The third [random + memory] machine keeps an "episodic memory" of the previously rejected rules and draws only from among the remaining possible rules.

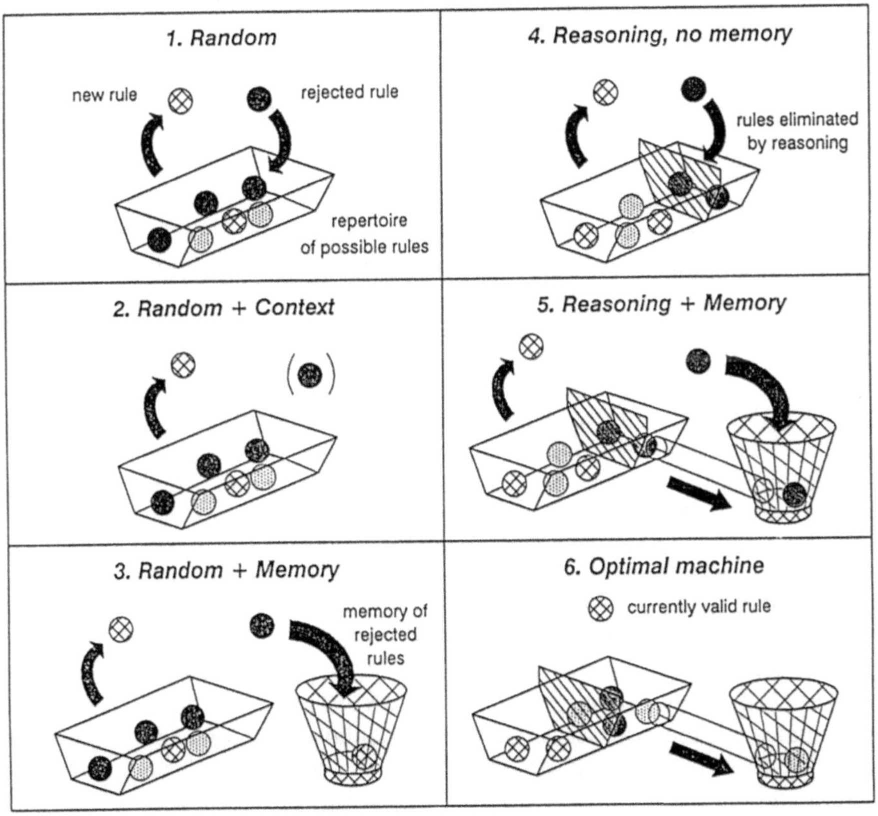

Fig. 6. Diagrammatic representation of the operation of six machines of increasing complexity learning the rules of the Wisconsin test by selection (taken from Dehaene and Changeux 1991)

The other three machines possess the additional faculty of rejecting rules by a type of tacit *"reasoning"* without having tested them overtly by trial and error. This is a very simple form of reasoning in the sense that, in the event of negative reinforcement, the machine eliminates all rules (in addition to that actually tested) which would lead to the same failure. The fourth machine [reasoning, no memory] utilizes reasoning in an extemporaneous manner without applying memory. The fifth machine [reasoning + memory] keeps in the memory a sketch of the previously rejected rules. Finally, the sixth, called "optimal," applies reason to the failures and memorizes the rules which have failed, but also applies reason to positive tests. Rules which have not given the same positive response are rejected and memorized as incorrect.

Comparison of the properties of these machines with the results of the Wisconsin card sorting test shows that this test does not allow all these properties to be tested. Of the three fundamental cognitive abilities of these machines – namely

1. ability to change the rule when punished,
2. memory of rules already tested and
3. *a priori* rejection of rules by reasoning – only the first is tested in a critical manner by the test.

A Formal Neuronal Architecture Able to Pass the Wisconsin Card Sorting Test

The model presented here (Dehaene and Changeux 1991) covers the major outlines of the architecture of the proposed formal organism in the case of tasks with delayed response, naturally with several major additions and modifications (Fig. 7).

The clusters of input neurons are more numerous since there are more categories and cue features involved in the test. Clusters of memory neurons are also present and receive projections from imput clusters with conservation of topography. There is competition between input clusters, with reciprocal inhibition so that only one feature is memorized for each dimension.

The memory neuron clusters project onto a new layer compared with the previous model. This layer is composed of neuron clusters which code for *"intentions"* of motor response that are distinct from the motor command itself. The model includes four clusters of intention neurons. Each code for the choice of a particular reference card, and activation of one of them excludes activation of the others. The intention is converted into an output command when an external "go" signal is received. Finally, like the previous model, this model includes clusters of rule-coding neurons which play a critical role in the performance of the test. Indeed, they modulate the connections between memory clusters and intention clusters according to a defined category (color, shape, number, etc.) and their activity varies with time during learning since each cluster which is active at a given moment inhibits the others. The organism uses them to test "hypotheses" of rules of behavior and selects a particular rule by interpreting a reinforcement signal. In fact, the network has been modelled in

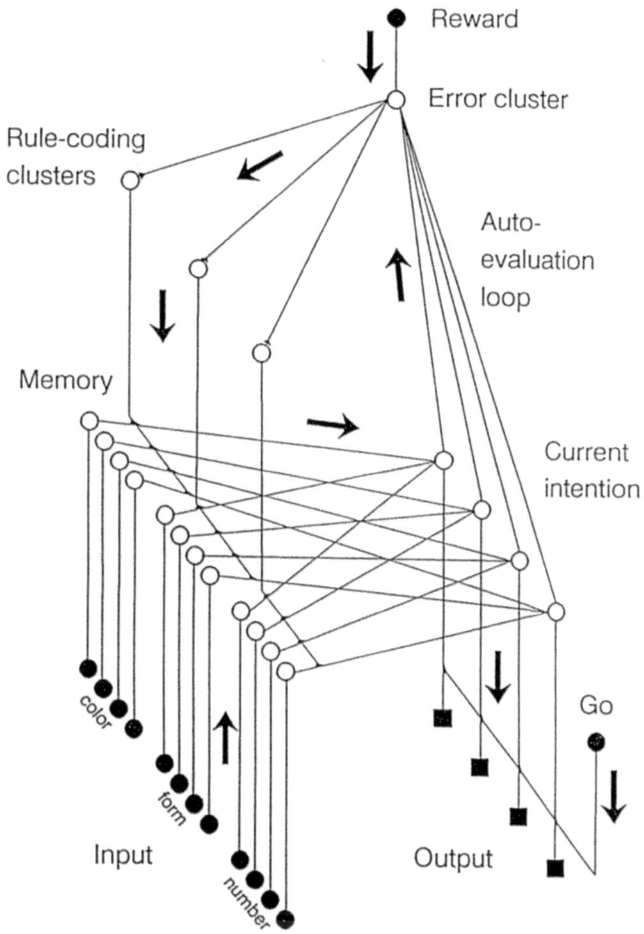

Fig. 7. Model of the role of the frontal cortex in the Wisconsin card sorting test. *Filled circles* represent the input clusters, *filled squares*, the output clusters, and *empty circles*, the internal clusters. Each *line* represents a bundle of synapses; auto-excitatory connections and lateral inhibitory connections within each assembly are not shown. On the input side, cards are coded along the dimensions of color, form, and number, and their features are stored in the short term by the self-excitatory connections of memory clusters. Memory clusters in turn activate the clusters defining the current intention for output. *Rule-coding clusters* modulate this transmission between *memory* clusters and *current intention* clusters, thus effectively deciding on the sorting rule. When the *go* unit is activated, activity coding for the current intention is in turn transmitted to output units. The subsequent entry of positive or negative reward (*top*) selects among the possible states of activity of the rule-coding layer, until the appropriate rule-coding cluster is activated. In the absence of an external reward, an auto-evaluation loop enables the system to reject rules autonomously, by evaluating the current intention with respect to memorized situations (taken from Dehaene and Changeux 1991).

such a way that each incorrect rule leads to punishment which, as in the previous model, destabilizes all the rule-coding neuron clusters so that they fluctuate in time and serve as the "generator of diversity."

Another novelty of the model is the differentiation of a *"cluster of error neurons"* which project and modulate the connections with rule-coding neuron clusters. The activity of error neurons is itself governed by reward signals so that a negative reward leads to short-term depression of excitatory connections in clusters of active rule-coding neurons. A molecular embodiment of this effect is allosteric regulation of postsynaptic receptor desensitization of the type described previously. This depression is spontaneously reversible and the speed of recovery is a crucial parameter which determines the memory range of the generator of diversity. If this speed is fast, the cluster of rule-coding neurons which has just been eliminated immediately enters again into the generator of diversity; it is a [random] machine. If the recovery speed is slow, a [random + context] machine is obtained which retains only the rule that has just been eliminated. Finally, when this speed is very slow, recovery extends over several consecutive tests and the network memorizes all the rules which have failed. It then behaves like a [random + memory] machine. The most original feature of the new model is the *"auto-evaluation loop"* which short-circuits the reward input from the exterior. This allows endogenous activation by intention clusters of error clusters, the efficacy of which is changed according to a classical Hebb's scheme. When a negative reward is received, the error neurons are activated and the connection linking intention clusters which are active at that moment with error clusters is reinforced. This intention is labelled as incorrect. Due to the persistence of activity in the error neurons, a new rule is tested within the rule-coding layer. This new rule is applied to the memorized features of the preceding cue, which produces a new distribution of intention cluster activity. If this distribution is identical to the previous one, the rule is rejected because the activity of the error cluster is maintained by potentiation of the intention/error connection, which prevents stabilization of the new rule. The "internal evaluation" of rules sequence is pursued until a correct rule is found.

Simulation of networks possessing auto-evaluation and memory shows a single trial percentage success rate much higher than that of the [random + memory] machine (98.4% versus 39.8%). Similarly, a network with an auto-evaluation loop but no memory is more successful than the [random + context] machine (Fig. 6).

Lesioning of the error cluster leads to slowing of learning and an increase in perseverations similar to those observed in frontal patients (see above). The inertia of the generator of diversity becomes very large. As in the case of the simple network, lesioning of rule-coding clusters interferes with the acquisition of a "systematic" rule of behavior. Lesioning of the auto-evaluation loop has no major qualitative effect on the behavior of the organism except for a loss of ability to reason, which significantly slows the learning process. However, it might offer a formal explanation of the "sociopathic" behavior resulting from ventromedian lesions of the frontal cortex (Damasio et al. 1990). Damasio et al. (1990) consider that this deficit is due to the inability to activate somatic states linked to the punishment or reward which the subject has experienced in association with specific social situations and which must be reactivated in connection with the anticipated result of a possible response. Injury to the intention/error connection might, according to our scheme, be the origin of this type of syndrome, evidently within a context both verbally and socially richer than that which served for modelling.

Conclusion

The two proposed formal models of the neuron network take into account the characteristic functional abilities of the prefrontal cortex: success in various delayed response tasks and in the Wisconsin card sorting test. They are based on principles of molecular, cellular and histological architecture that are plausible at the neurobiological level. These models are extremely simple and might even appear simplistic to cerebral cortex specialists. Nevertheless, they provide several original and specific predictions able to delineate novel experimental tests. One bears on the existence of "rule-coding neurons," the activity state of which varies randomly during the learning period until a rule of behavior is selected. Another concerns the mechanism, or mechanisms, of reinforcement by "error neurons." On the one hand their activity is regulated by that of neurons coding for motor intentions, and on the other hand they exert a regulatory action on rule-coding neuron clusters.

At a more general level, the induction of rule by trial and error followed by selection integrates perfectly with evolutionary epistemology (Changeux and Dehaene 1989; Changeux et al. 1973; Changeux 1983; Popper 1966; Campbell 1974; Darwin 1859; Poincare 1913; Dehaene and Changeux 1991; Edelman 1978) and illustrates the concept of "mental Darwinism" (Changeux and Dehaene 1989; see also Campbell 1974). In this context, clusters of rule-coding neurons would constitute the "generator of diversity." Memorization by selection may be considered as a homologue of "amplification" since the organism will re-use the memorized trace repeatedly in its subsequent behaviors.

The models also illustrate the precise contribution of hierarchical levels of network architecture to defined behaviors and particularly:

1. the ability to generalize a rule acquired for a particular cue to an entire class of cues, or *systematicity* (Dehaene and Changeux 1989; Fodor and Pylyshyn 1988);
2. the ability to "memorize" rules which have already been tested on the outside world; and
3. the ability to evaluate new rules in a tacit manner by internal auto-evaluation which may be taken as a very simple form of "reasoning" (Dehaene and Changeux 1991).

Finally, these models and their simulation show how some elementary components of the network (e.g. allosteric receptors, synaptic triads) can introduce constraints into higher cognitive functions ("bottom-up" regulation). They also illustrate how a global process of interaction with the outside world, such as reward or reinforcement, can govern regulation at a more elementary level, such as the regulation of conformational transitions of allosteric receptors ("top-down" regulation). Last of all, they offer a specific illustration of the interdependence between levels of organization which confer structural coherence and functional integration on the system.

References

Amit DJ (1989) Modeling brain functions. Cambridge University Press, Cambridge
Batuev AS, Orlov AA, Pirogov AA (1981) Short-term spatio-temporal memory and cortical unit reactions in the monkey. Acta Physiol Hung. 58: 207–216
Campbell DT (1974) Evolutionary epistomology.In: Schilpp PA (ed). The philosophy of Karl Popper. La Salle, Open Court
Changeux JP (1983) L'Homme neuronal. Fayard, Paris; English edition: Neuronal man, Pantheon New York
Changeux JP (1988) Molécule et mémoire. Bedou, Gordes
Changeux JP, Courrege P, Danchin A (1973) A theory of the epigenesis of neural networks by selective stabilization of synapses. Proc Natl Acad Sci USA 70: 2974–2978
Changeux JP, Dehaene S (1989) Neuronal models of cognitive functions. Cognition 33: 63–109
Changeux JP, Heidmann T (1987) Allosteric receptors and molecular models of learning. In: Edelman G, Gall WE, Cowan WM (eds) Synaptic Function. John Wiley, New York, pp 549–601
Damasio AR, Tranel D, Damasio H (1990) Individuals with sociopathic behavior caused by frontal damage fail to respond autonomically to social stimuli. Behav Brain Res 41: 81–94
Darwin C (1859) On the origin of species. London: Murray
Dehaene S, Changeux JP (1989) A simple model of prefrontal cortex function in delayed-response tasks, J Cognitive Neurosci 1: 244–261
Dehaene S, Changeux JP (1991) The Wisconsin card sorting test: theoretical analysis and simulation of a reasoning task in a model neuronal network. Cerebral Cortex, in press
Dehaene S, Changeux JP, Nadal JP (1987) Neural networks that learn temporal sequences by selection. Proc Natl Acad Sci USA 84: 2727–2731
Diamond A (1985) The development of the ability to use recall to guide action, as indicated by infant's performance on AB. Child Develop 56: 868–883
Diamond A (1988) Differences between adult and infant cognition: is the crucial variable presence or absence of language? In: Weiskrantz L (ed.) Thought without language. Clarendon Press, Oxford
Edelman GM (1978) Group selection and phasic reentrant signaling: a theory of higher brain function. In: Edelman GM, Mountcastle VB (eds) The mindful brain: cortical organization and the group-selective theory of high brain function. MIT Press, Cambridge, pp. 51–100
Fodor JA, Pylyshyn ZW (1988) Connectionism and cognitive architecture: A critical analysis. Cognition 28: 3–71
Fuster JM (1973) Unit activity in prefrontal cortex during delayed-response performance: Neuronal correlates of transient memory. J Neurophysiol 36: 61–78
Fuster JM (1984) Electrophysiology of the prefrontal cortex. Trends Neurosci 1: 408–414
Fuster JM (1989) The prefrontal cortex (2nd edition). Raven Press, New York
Fuxe K, Agnati L (1991) Volume transmission in the brain: new aspects for electrical and chemical communication. Raven, New York
Goldman-Rakic P (1987) Circuitry of the primate prefrontal cortex and the regulation of behavior by representational knowledge. In: Mountcastle V, Plum KF (eds) The nervous system: Higher functions of the brain, Vol. 5, Handbook of Physiology. American Physiological Society, Washington, DC:
Grant DA, Berg EA (1948) A behavioral analysis of degree of reinforcement and ease of shifting to new responses in a Weigl-type card-sorting problem. J Exp Psych 38: 404–411
Heidmann T, Changeux JP (1982) Un modèle moléculaire de régulation d'efficacité d'un, synapse chimique au niveau postsynaptique. C R Acad Sci Paris, série 3, 295: 665–670

Jacob F (1970) La logique du vivant. Gallimard, Paris

Kubota K, Niki H (1971) Prefrontal cortical unit activity and delayed alternation performance in monkeys. J Neurophysiol 34: 337–347

Kubota K, Komatsu H (1985) Neuron activities of monkey prefrontal cortex during the learning of visual discrimination tasks with go-no go performances. Neurosci. Res. 3: 106–129

Marr D (1982) Vision. Freeman, San Fransisco

McCulloch WS, Pitts W (1943) A logical calculus of the ideas imminent in nervous activity. Bull Math Biophys 5: 115–137

Milner B (1963) Effects of brain lesions on card sorting. Arch Neurol 9: 90–100

Mountcastle VB (1978) An organizing principle for cerebral function: the unit module and the distributed system. In: Edelman GM, Mountcastle VB (eds) The mindful brain. The MIT Press, Cambridge

Nauta WJH (1971) The problem of the frontal lobe: a reinterpretation. J Psych Res 8: 167–187

Newell A (1982) The knowledge level. Artificial Intelligence 18: 87–127

Niki H (1974) Differential activity of prefrontal units during right and left delayed response trials. Brain Res 70: 346–349

Niki H (1975) Differential activity of prefrontal units during right and left delayed response trials. In: Kawai M, Ehara A and Kawamura S (eds) Symposia of the Fifth Congress of International Prematological Society, Kondos, Japan Science Press, Tokyo, pp 475–486

Piaget J (1954) The construction of reality in the child. Basic Books, New York

Poincare H (1913) Science et Méthode. Flammarion, Paris

Popper KR (1963) Conjectures and refutations. Basic Books, New York, 1963

Popper KR (1966) Of clouds and clocks: an approach to the problem of rationality and the freedom of man. Washington University Press, St. Louis, Missouri

Rosenkilde E, Bauer CE, Fuster JM (1981) Single cell activity in ventral prefrontal cortex of behaving monkeys. Brain Res 209: 375–394

Shallice T (1982) Specific impairments of planning. Phil Trans Royal Soc London B, 298: 199–209

Teuber HL (1964) The riddle of frontal lobe function in man. In: Warren JM, Akert K (eds) The frontal granular cortex and behavior. McGraw-Hill, New York, pp. 410–477

Teuber HL (1972) Unity and diversity of frontal lobe functions. Acta Neurobiol Exp (Warsz.) 32: 615–656

Watanabe M (1986a) Prefrontal unit activity during delayed conditional Go/No-Go discrimination in the monkey. II. Relation to Go and No-Go responses. Brain Res 382: 15–27

Watanabe M (1986b) Prefrontal unit activity during delayed conditional discriminations in the monkey. Brain Res. 225: 51–65

Exploring Memory in the Aging Brain from the Perspective of Recent Split Brain Studies

L. L. Le Sueur and *M. S. Gazzaniga*

Introduction

Few individuals over the age of 60 will deny that their memory is not as good as it used to be. Actually, by the age of 45 most of us will have already begun to notice that we cannot learn or recollect as easily or as accurately as we used to. Memory loss may be the most common complaint among middle-aged people and senior citizens, second perhaps only to the loss of normal health. It is, naturally, not the least bit comforting to have the phenomenon referred to as "benign senescent forgetfulness" (Kral 1972). Despite these complaints or the denial of symptoms, the fact is that human memory declines with age. We have only to look at the normative data for the Wechsler Memory Scale-Revised to see that this decline in memory is not merely anecdotal, but true for the population at large. In addition, there is a marked decrease in retention of learned materials for the elderly (see Fig. 1; Wechsler 1987).

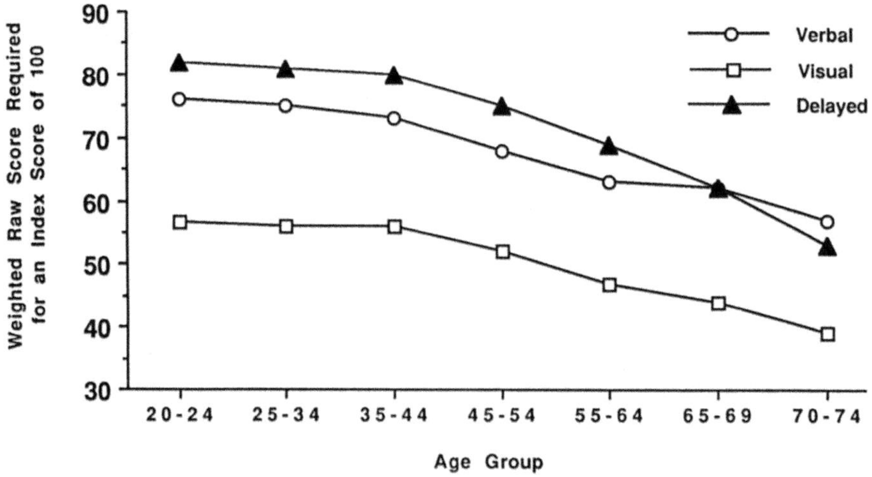

Fig. 1. Wechsler Memory Scale-Revised. This graph is intended to display the decline pattern within each of these subtests only. It is not reasonable to compare raw scores across the Verbal, Visual, and Delayed Recall subtests. Also, note the more precipitous decline in Delayed Recall scores, indicating for the normal population an increasing loss in retention capacity over time (Adapted from Wechsler 1987)

There is much experimental evidence to support the painful truth that we all lose our memories as we age. One of the more robust of these findings is that recall and recognition performances decouple over time. Rather small differences are repeatedly found for recognition tasks across age groups, but increasingly larger ones occur for recall tasks (e.g., Shonfield and Robertson 1966). However, when encoding strategies such as imagery are employed, recall performance in the elderly improves (e.g., Hulicka and Grossman 1967; Poon et al. 1980). Such findings suggest a pattern of loss in retrieval rather than encoding processes. This idea is further supported by findings that semantic memory is preserved with age (e.g., Howard 1983), by observations that cued recall declines more slowly than free recall (e.g., Hultsch 1975), and by reports from normal older subjects of more frequent experiences with "tip-of-the-tongue" phenomena (Burke and Harrold 1988) that the elusive memories usually emerge spontaneously.

Like the normal aging population, patients with Alzheimer's dementia perform better on tests that tap overlearned knowledge and on tests that are familiar, concrete, and not speed-dependent. However, for Alzheimer's patients, both recall and recognition are compromised, and their memory losses go beyond difficulties with name finding. Their everyday problems include the inability to keep track of time, days, months, news, finances, or plots of TV shows. These patients often have difficulty remembering phone numbers, typing, or finding their way around once familiar settings. Moreover, verbal deficits are far more profound for Alzheimer's patients. Finally, although Alzheimer's patients often recount events in a somewhat creative fashion, i.e., "confabulate" what they cannot accurately remember, they will occasionally exhibit "windows of clarity" in which their responses are lucid and accurate.

These "windows of clarity" in Alzheimer's patients present an intriguing paradox: How can the memory of a progressively dementing mind have occasional moments of relatively cogent thought process, with intact facts and logical structure? Some unique insights into this conundrum may exist in data from split brain research.

Memory in Split Brain Patients

In an attempt to control intractable epilepsy, the corpus callosum – a dense bundle of nerve fibers connecting the two cerebral hemispheres – is sometimes surgically severed. Participants in this radical procedure rarely manifest many noticeable everyday behavioral changes, except that in most cases their seizures subside. However, closer laboratory scrutiny has revealed certain patterns of functional hemispheric specialization (e.g., Gazzaniga and Ledoux 1978; Sperry et al. 1969).

Early memory studies with split brain patients suggested a short-term memory deficit, but reports varied a great deal from patient to patient. Memory loss in these patients was observed as significantly greater than for other epileptic patients (Zaidel and Sperry 1974). However, a better comparison would be between the pre- and post-operative memory performances for each patient.

When this more desirable design was first employed, memory actually appeared to improve considerably after surgery (Ledoux et al. 1977). The important caveat to this finding is that the surgery on the patient reported in this study had most likely spared callosal fibers in the posterior region of the callosum.

By contrast, JW, a split brain patient whose callostomy is complete, has shown a marked deficit in recall relative to fairly stable recognition scores following brain bisection (Gazzaniga 1987). This pattern in recall and recognition losses is similar to that found in the normal aging population. More recently Phelps et al. (submitted for publication) conducted an experiment to explore the possibility that split brain patients with anterior versus posterior partial sections show differential losses in memory function. Their data suggest that severing the connections between the anterior regions actually improved memory performance in two of the patients. However, when the posterior cortical regions were cut, a significant loss in recall performance for both verbal and visual tasks was documented for both of the patients. It is important to note here that the hippocampal commissure is included in the posterior transection.

Structural damage to the hippocampus has become strongly implicated in Alzheimer's disease. Although cortical structures are eventually globally affected, the entorhinal cortex (Brodman's area 28) of the ventromedial temporal lobe is consistently a focus of pathology in Alzheimer's disease. This region gives rise to axons bidirectionally interconnecting the hippocampal formation with remaining cortex (e.g., van Hoesen et al. 1991). Although hippocampal disconnection may not be enough to result in the full spectrum of memory losses seen in Alzheimer's patients, it may still describe an important aspect of memory loss for both Alzheimer's patients and normally aging individuals. Undermining the posterior connections, either by surgery or pathology, may prevent the kind of encoding processes that are more important for recall than recognition (such as integrating, elaborating, associating, and reflecting; see Phelps et al. submitted) processes that tend to diminish with normal aging and Alzheimer's disease.

But in the Alzheimer's patient we are still left with the paradoxial problem of a progressive disease that can still allow lucid moments or "windows of clarity." These patients have severe immediate and delayed recall problems. Could it be that their increasing state of confusion is a product not only of poor memories, but also of weakening interpretations of what could only be called a noisy data base from which they try to evaluate their experiences? In the same light, could their windows of clarity represent brief moments when an "interpreter" is still capable of managing short term memories that happen to be accurate for a given event?

The Interpreter

In everyday experience, if simple, uncomplicated retrieval fails, there is a tendency in all of us to weave together the available data to fashion a complete "memory". Actually, as Bartlett (1932) pointed out, "Remembering is not ... fixed, ... [but] an imaginative reconstruction ...." To accomplish this feat, we rely not only on the data, but also on various logics, such as previous knowledge, temporal sequence, schemata, etc. Whenever the data are faulty or inadequate, we can generally recognize these recollections as reconstructions. But if the data are poor or inadequate, or the reconstruction is poorly framed, illogical, or unacceptable, we generally agree this is confabulation and pathological, and suspect there is no "self-awareness." In either instance, of normal reconstruction or pathological confabulation, there seems to be at least an executive attempt at imposing order on the data. This ordering mechanism has been referred to by Gazzaniga (1985, 1987, 1989) as the "interpreter."

Evidence from split brain studies has suggested that our apparently unified mental activities are dissociable. In one interesting "simultaneous concept" experiment, a series of picture pairs was shown to a split brain patient, one picture to the left and one to the right hemisphere (e.g., a chicken and a snow scene, respectively). From an array of pictures displayed in front of the patient, each hemisphere would correctly select an object associated with the picture displayed to it (e.g., a claw and a shovel). However, while the left hemisphere could accurately explain the choice of the claw for the chicken, it was also compelled to explain away the shovel by saying "you need a shovel to clean out the chicken shed." Another typical example is the patient "going for a Coke" in explanation for leaving the testing area after the word "walk" was flashed to the right hemisphere. Phelps et al. (submitted for publication) have found experimental evidence from two patients with essentially complete commissurotomies that the interpretive system may be a left hemisphere function. After viewing stories scene by scene, the patients responded in a recognition task to lateralized targets consisting of scenes actually viewed, scenes related but not viewed, and foils. If an interpreter is operating, then more than chance false positives would be expected for the related scenes. The related scenes did draw significantly more false positives, and more when viewed by the left hemisphere than the right.

The notion of an interpretive system that manipulates the data of separate and specialized processing subsystems is based on numbers of these observations. The processing subsystems or modules are capable of executing their functions in parallel, without the conscious awareness of the interpretive system. The interpreter takes the modular output data and fits them into an existing mental schema. It is intrinsically linked to the capacities of making logical, causal inferences, and drawing hypotheses. The interpreter affords us "self"-awareness and the "knowing that you know" experience.

It seems reasonable to expect, though, that an interpreter will function best when demands on it are not excessive, even for the normal and young population, but especially if it is declining. However, if the interpreter is weakened in some way, by fatigue, stress, disease, or a limited repertoire of applicable frameworks, we could also expect recollection to appear poor, incoherent, or confabulatory. For Alzheimer's patients confabulating their responses, this interpreter is still operating, albeit imperfectly and sometimes without full self-awareness, on a data base limited by the deteriorating hippocampal structures, which is to say the recall mechanism. The result is an output of inaccurate or inadequate details loosely woven together by an inadequate or innappropriate schematic structure. In contrast, whenever the details happen to be adequate or accurate, and the framework sufficient or appropriate, the output appears as a "window of clarity."

Conclusion

The above results from split brain studies give rise to two hypotheses about memory changes in normal aging and Alzheimer's disease. One is that some integration across the hippocampal hemispheres appears to be necessary in order for successful encoding and/or retrieval to occur. This structure is believed to mediate the consolidation of long-term memory processes (Scoville and Milner 1957; Squire and Zola-Morgan 1983; Victor and Agamanolis 1990). It has also been speculated that the hippocampus may be important for integrative, reflective, or associational processes (e.g., Hebb 1961). Indeed, successful integration of all sensory/motor/linguistic input would seem to provide the highest probability that an event could be established as a long-term memory and thus be reactivated spontaneously.

The decline with age of hippocampal integration could provide a parsimonious explanation for many of the deficits described in the literature for normal aging processes, as well as those observed in Alzheimer's. A decline in the encoding/retrieval powers of the integrating mechanism would certainly precipitate naming and word finding and "tip-of-the-tongue" phenomena. Complex materials would necessarily be more difficult to comprehend with this model because they would challenge a limited integration system. Novel and ambiguous materials would also make high integrative demands on a deficient hippocampus. Clearly Alzheimer's patients suffer losses beyond those of normals, but the hippocampus may be the common structural denominator.

The second hypothesis offered toward understanding memory loss in Alzheimer's disease is that interpretive mechanisms underlying conscious experiences may be altered. Thus, the manipulation of the data base, however rich, varied, or inadequate it may be, and the capacity to order experience according to established knowledge structures, or belief systems, seems to be necessary for a fully functioning memory system. In contrast, this mechanism appears to remain fairly well-preserved in normal aging.

In normal aging, a weakened hippocampus would result in a limited data base and integrative powers, so that processing unusual material becomes restricted and retrieval is thus more difficult. In Alzheimer's patients, the integrated data base has also become impoverished, but whenever the weakening interpreter attempts to organize the faulty data base, confabulation results. The paradoxical window of clarity can occur because the data base is not always inadequate, and the interpreter will continue to order the data as best it can. In cases where overlearned and uncomplicated schemata are all that are demanded of the patient (e.g., introductions, greetings, describing what is desired for breakfast, etc.), the combination with an adequate but reduced data base may be sufficient for momentary lucidity.

References

Bartlett FC (1932) Remembering. Cambridge University Press, Cambridge

Burke DM, Harrold RM (1988) Automatic and effortful semantic processes in old age: experimental and naturalistic approaches. In: Burke DM, Light LL (eds) Language, memory, and aging. Cambridge University Press, Cambridge, pp 100–116

Gazzaniga MS (1985) The social brain. Basic, New York

Gazzaniga MS (1987) Cognitive and neurologic aspects of hemispheric disconnection in the human brain. In: Magistretti (ed) Discussions in neuroscience, vol. 4. Fondation pour l'Etude

Gazzaniga MS (1989) Organization of the human brain. Science 245: 947–952

Gazzaniga MS, Ledoux JE (1978) The integrated mind. Plenum, New York

Hebb DO (1961) Distinctive features of in the higher animal In: Delafresnaye JF (ed) Brain mechanisms and learning. Blackwell, London

Howard DV (1983) The effects of aging and degree of association on the semantic priming of lexical decisions. Exp Aging Res 9: 145–151

Hulicka IM, Grossman JL (1967) Age-group comparisons for the use of mediators in paired associate learning. J Gerontol 22: 46–51

Hultsch D (1975) Adult age differences in retrieval: Trace-dependent and cue-dependent forgetting. Dev Psychol 11: 197–201

Kral VA (1972) Senescent forgetfulness, benign and malignant. Can Med Assoc J 86: 257–260

LeDoux JE, Risse G, Springer SP, Wilson DH, Gazzaniga MS (1977) Cognition and commissurotomy. Brain 100: 87–104

Poon LW, Walsh-Sweney L, Fozard JL (1980) Memory skill training for the elderly: salient issues on the use of imagery mnemonics. In: Poon LS, Fozard JL, Cermak LS, Arenberg D, Thompson LW (eds) New directions in memory and aging. Erlbaum, Hillsdale, pp 461–484

Scoville WB, Milner B (1957) Loss of recent memory after bilateral hippocampal lesions. J Neurol Neurosurg Psychiatry 20: 11–21

Shonfield D, Robertson BA (1966) Memory storage and aging. Can J Psychol 20: 228–236

Sperry RW, Gazzaniga MS, Bogen JE (1969) Interhemispheric relationships; the neocortical commissures; syndromes of hemisphere disconnection. In: Vinkin PJ, Bruyen GW (eds.) Handbook of clinical neurology, vol 4. North Holland, Amsterdam

Squire LR, Zola-Morgan S (1983) Memory: brain systems and behavior. Trends Neurosci 11: 170–175

Van Hoesen GW, Hyman BT, Damasio AR (1991) Entorhinal cortex pathology in Alzheimer's disease. Hippocampus 1: 1–8

Victor M, Agamanolis D (1990) Amnesia due to lesions confined to the hippocampus: a clinical pathologic study. J Cogn Neurosci 2: 246–257

Wechsler D (1987) Wechsler Memory Scale-Revised. Harcourt Brace Jovanovich, San Antonio

Zaidel E, Sperry RW (1974) Memory impairment following commissurotomy in man. Brain 97: 263–272

Alzheimer's Disease: Disruption of Mind-Brain Relations

S. I. Rapoport

Summary

Measurements of failing cognitive abilities in individual Alzheimer's disease (AD) patients, using specific neuropsychological tests, when correlated with measurements of regional brain energy metabolism and blood flow, using the *in vivo* imaging technique of positron emission tomography (PET), indicate that AD disrupts the relations of mind with brain and can be used to examine the nature of these relations. AD patient's demonstrate correlations between failing visuospatial and linguistic abilities, and reductions in "resting state" metabolism in the right and left hemispheres, respectively; between failing verbal fluency and attention and reduced metabolism in the prefrontal association cortex; and between failing arithmetic, verbal comprehension, drawing and immediate memory for visuospatial and reduced metabolism in the parietal association cortex. Fewer significant correlations between pairs of metabolic rates in the parietal and frontal cortices of the AD brain further reflect functional disintegration of these domains.

Nevertheless, blood flow studies in mildly-moderately demented AD patients show that occipitotemporal association regions, part of the metabolically dysfunctional network subserving object recognition, can be normally activated in patients performing a face matching task. Thus, an early and initially reversible failure of synaptic transmission likely precedes and accounts for disruption of some mind-brain relations in the initial phase of AD.

Introduction

An important question in neurophilosophy is whether, by logical reasoning or direct experimental observation, the operations of the human mind – cognition, behavior and emotion – can be reduced (described or mapped) by the structure and function of the human brain. Some philosophers believe that all aspects of mind eventually will be reducible to the algorithms carried out by neurons and neuronal nets within the brain (Churchland 1986; Changeux, in Changeux and Connes 1989), whereas others argue that some aspects, including mathematical insight, creativity, and goal setting, are not reducible (Penrose 1989; Connes, in Changeux and Connes 1989).

The extent to which mind can be reduced to brain is the subject of intense research in the neurobiological and cognitive sciences, and becomes more and more detailed as new information is gathered. Bertrand Russell's contention (Gove 1966) that "every advance in knowledge robs philosophy of some problems it formerly had" is proving true daily. In this discussion, I shall consider how studies of a neurodegenerative disorder, Alzheimer's disease (AD), using a new in vivo brain imaging technique – positron emission tomography (PET) – and detailed neuropsychological tests, have helped and promise to help to define the extent and nature of the map of mind onto brain.

AD, the fourth most common cause of death in the United States (Katzman 1976), is a human neurodegenerative disorder whose prevalence increases with age. AD affects approximately 17% of individuals over the age of 65 (Cummings 1989). It often becomes evident as defects in memory, attention and planning, followed in variable sequence by failure of language and visuospatial functions and a general deterioration of "global intelligence," resulting in a dementia syndrome (Haxby et al. 1988). AD has a variable but steady rate of decline in different patients, and results in death after a course of about 8 years (Heston et al. 1981).

The postmortem AD brain has a number of features which allow for a diagnosis of AD with clinical "certainty" (McKhann et al. 1984). However, not one of these features is unique to AD, and as yet a specific biological marker for AD does not exist. Senile (neuritic) plaques, degenerating terminals of long axons, are evident in the AD brain (Terry and Wisniewski 1972). These plaques frequently contain beta-amyloid, a 4.2 kD peptide that is a product of amyloid precursor proteins which are produced in part within neuronal cell bodies (Palmert et al. 1988; Chou et al. 1990). It has been argued that overproduction of this beta-amyloid peptide is pathogenic in AD. Neurofibrillary tangles (consisting of paired helical filaments of abnormally phosphorylated fibrillary proteins; Grundke-Iqbal et al. 1986) are found in large numbers in neurons of the AD brain, suggesting also that abnormal phosphorylation or dephosphorylation of intracellular neuronal fiber elements contributes to disease pathology.

AD is a common phenotype which may have any of several possible causes. It can occur in sporadic (nonfamilial) or in familial (heritable) forms (Heston et al. 1981; Rapoport et al. 1991b). In the latter form, chromosome 21 has been implicated in some families (St George-Hyslop et al. 1987), chromosome 19 in orthers (Roses et al. 1990). A role for genetic factors is supported by evidence that AD degeneration occurs in all subjects with Down syndrome (trisomy 21) over the age of 35 (Wisniewski et al. 1985; Schapiro et al. 1990). On the other hand, sporadic AD may be a stochastic event predicted by the character of the individual's genome, as influenced by nongenetic, including environmental, factors (Rapoport 1988; Rapoport et al. 1991b).

Measurements of failing cognitive abilities in individual AD patients, using specific neuropsychological tests, were correlated with measurements of regional brain energy metabolism and blood flow, using the imaging technique of PET. The results provide firm evidence that AD disrupts the relations of mind with brain. Thus, AD can be used to examine the nature and extent of these relations in humans. This will be the subject of my later discussion.

The anatomic location of metabolic abnormalities in living AD patients corresponds to the location of neuropathology in the postmortem AD brain (Rapoport 1990a); both distributions demonstrate selective involvement of the association neocortices and related non-neocortical regions. To understand this selectivity, it should be noted that each primary sensory or motor neocortical area is connected reciprocally with first-order and second-order modality-specific association cortical areas, which in turn are connected with polymodal association areas chiefly within the intraparietal sulcus, the parietotemporal cortex, and prefrontal cortex (Pandya and Seltzer 1982; Goldman-Rakic 1988; Rapoport 1990b). Association neocortices also are connected reciprocally with many regions within the telencephalon and diencephalon, including the amygdaloid complex, entorhinal cortex, nucleus basalis of Meynert, cingulate gyrus and hippocampal formation; thalamic nuclei and basal ganglia; and mesencephalic-midbrain catecholaminergic nuclei. Long intracortical fibers originate principally from pyramidal neurons within layers III and V–VI of the neocortex, whereas corticofugal fibers originate mainly from pyramidal neurons in layer V (Wise and Jones 1977).

AD neuropathology, particularly neurofibrillary tangles and cell loss, is selective to pyramidal cells of layers III and V of the association neocortices (Lewis et al. 1987), and to regions within the hippocampal formation, amydaloid complex and entorhinal cortex, and within subcortical cholinergic and catecholaminergic nuclei, which are intimately connected with the neocortex (reviewed by Rapoport 1988, 1989, 1990a, b). With the exception of the brainstem catecholaminergic nuclei, most of these regions belong to a telencephalic association "system" which is thought to have expanded and differentiated during evolution of higher primates, particularly of hominids.

As far as is known, AD is a distinctly human disease; although limited AD pathology (e.g., senile plaques) has been reported in brains of nonhuman primates (Selkoe et al. 1987), there is no evidence for an equivalent, full-blown disease process in any but the human species. This fact, and the colocalization of metabolic dysfunction and AD neuropathology in the association system, suggest that AD is a phylogenic disease, that is, a disease which involves brain regions on the basis of their coherent and genetically based elaboration during evolution (Hughlings Jackson 1884, in Taylor 1931; Roofe and Matzke 1968; Rapoport 1988, 1989, 1990b).

If this hypothesis is correct, an understanding of the genetic changes that promoted brain evolution of the telencephalic association system in higher primates should provide information about the genetic basis of its vulnerability to AD degeneration, and vice versa. Furthermore, statistically significant correlations between patterns of failing higher cognitive function and patterns of failing brain metabolism in AD patients (see below) should be consistent with how these patterns and related neuroanatomy coevolved during primate phylogeny.

Consistent with this prediction is evidence that many analogous, higher order cognitive functions are not discontinuous between nonhuman primates and humans, and that they can be identified with homologous brain networks in the different species (Rapoport 1990b). For example, common brain network features have been proposed to subserve a universal grammar in the great apes, implying that these species can command parts of this common structure for elementary symbolic logic and communication (Chomsky 1980; Gallup 1977; Menzel 1973; Premack and Premack 1972). Furthermore, common anatomic and functional features of cortical visual pathways subserving spatial as compared with object discrimination in the rhesus monkey and humans imply homology and continuity between the two species (Mishkin et al. 1983; Kaas 1989; Haxby et al. 1991; and see below).

Profiles of Cognitive Failure in AD Patients

To examine the deteriorating relations between mind and brain in AD, AD patients of different dementia severities were chosen, defined by their score on a Mini-Mental State Examination: mild, score = 22–30; moderate, score = 11–21; severe, score = 0–10 (Folstein et al. 1975). They, as well as age-matched controls, were administered a battery of neuropsychological tests designed to quantify various parameters of cognition; the results of these tests are listed in Table 1 (Haxby et al. 1988). Their dementia was further characterized by the Mattis Dementia Rating Scale (Mattis 1976) and the Wechsler Adult Intelligence Scale (WAIS; Wechsler 1955). The WAIS was scored to yield a full-scale intelligence quotient, as well as factor deviation quotients (DQ; Tellegen and Briggs 1967) that summarize performance on tests of verbal fund of knowledge (Verbal Comprehension), immediate verbal memory span and arithmetic (Memory and Freedom from Distractibility), and visuospatial construction (Perceptual Organization). The ability to commit new information to memory was measured with the story recall and visual reproduction subtests of the Wechsler Memory Scale (Wechsler 1945) with delayed recall (Russell 1975).

The ability to maintain attention to complex or shifting sets was measured with the Stroop Color Word Test (Golden 1978) and Trail B of the Trailmaking Test (Reitan 1958). Trail A of the Trailmaking Test measured attention to a simpler set. Planning and foresight were considered with the Porteus Maze Test (Porteus 1965). Language was analyzed using two tests. Comprehension of syntactic relations in single sentences was tested using Whitehouses's Syntax Comprehension Test (Haxby et al. 1985). Verbal fluency was tested using the Controlled Word Association (FAS) test (Benton 1973). Visuospatial construction was tested using the Extended Range Drawing Test, a measure of the ability to copy geometric figures of varying complexity (Haxby et al. 1985). Immediate memory span for visuospatial lotation was measured with a block tapping test patterned after Corsi's test (Milner 1971) with a 12-block array.

As illustrated in Table 1, moderately demented patients had lower scores than controls on all of the cognitive tests that were administered ($p < 0.05$). In contrast, mildly demented patients had significantly lower scores only on the Mattis Dementia Scale (confirming their dementia), on measures of visual recent memory (Wechsler Memory Scale), attention to a complex set (Trail-

Table 1. Cognitive profiles for Alzheimer's disease patients with mild or moderate dementia. Neuropsychological tests are arranged according to the general brain function that they are considered to evaluate (data from Haxby et al. 1988)[a]

Test administered	Controls (N = 19–29)	Mild AD (N = 10)	Moderate AD (N = 14)
Summary measures			
Mattis dementia scale	142 ± 2	132 ± 6*	110 ± 17**
WAIS full scale IQ	125 ± 10	117 ± 8	90 ± 19**
WAIS deviation quotients			
Verbal comprehension	129 ± 10	122 ± 9	99 ± 19**
Memory and distractibility	116 ± 14	114 ± 9	90 ± 16**
Perceptual organization	119 ± 13	108 ± 13	83 ± 25**
Memory (Wechsler memory scale)			
Immediate story recall	22 ± 5	11 ± 5**	5 ± 3**
Immediate visual reproduction	10 ± 3	6 ± 5*	1.4 ± 1.8**
Delayed story recall	17 ± 5	3 ± 4**	0.7 ± 1.1**
Delayed visual reproduction	7 ± 3	1 ± 1**	0.3 ± 0.7**
Attention and planning			
Trailmaking (trail A, s)	40 ± 17	54 ± 30	153 ± 98**
Trailmaking (trail B, s)	84 ± 38	202 ± 159*	434 ± 134**
Stroop color-word interference – no corr.	37 ± 9	24 ± 8*	12 ± 8**
Porteus mazes (test ages)	15.4 ± 1.5	12.5 ± 3.9*	7.5 ± 3.7**
Language			
Syntax comprehension	24 ± 3	23 ± 2	17 ± 5**
Controlled word association (FAS)	42 ± 15	31 ± 8	23 ± 12**
Visuospatial function			
Extended range drawing	21 ± 2	19 ± 4	13 ± 5**
Block tapping span	29 ± 6	25 ± 4	19 ± 7**

[a] Mean ± differs from control mean. *$p < 0.05$; **$p < 0.001$

making B and Stroop Color Interference Task), and planning (Porteus Mazes). In fact, mean neuropsychological deficits in mildly demented patients were limited to marked memory impairment and to a milder impairment on tests that require sustained attention to complex or shifting tests, whereas performance on most tests of focal neuropsychological functions was not significantly impaired. Moderately demented patients performed worse than mildly demented patients on all measures except those of delayed verbal memory, delayed visual memory and verbal fluency (Haxby et al. 1988).

Profiles of Failing "Resting State" Brain Metabolism in AD

Positron Emission Tomography for the Study of Brain Metabolism and Blood Flow

Roy and Sherrington (1890) suggested that regional cerebral blood flow (rCBF) is coupled to chemical products of cerebral metabolism, allowing flow to vary locally "in correspondence with local variations of functional activity." Since their observation, attempts have been made to measure rCBF to quantify local neuronal activity. Furthermore, as glucose is the major substrate for brain oxidative metabolism, the regional cerebral metabolic rates for glucose ($rCMR_{glc}$), as well as O_2 ($rCMRO_2$), also have proven useful in this regard.

In the last decade, positron emission tomography (PET) has made it possible to quantify rCBF, $rCMR_{glc}$ and $rCMRO_2$ in brains of awake humans, and to localize these rates with spatial resolutions approaching 5 mm (Huang et al. 1980; Alpert et al. 1984; Horwitz 1990). In the PET procedure, a positron-emitting compound which is administered systemically is taken up by brain, where it releases positrons (positively charged electrons). These collide with electrons and are annihilated, releasing two gamma rays at 180° to each other. A ring of radiation detectors surrounding the head can identify, by coincidence counting with appropriate reconstruction algorithms, the quantity and location of radioactivity within the brain.

To measure $rCMR_{glc}$, we inject the positron emitting tracer ^{18}F-2-deoxy-D-glucose (^{18}FDG; radioactive half-life of ^{18}F is 110 min) intravenously into subjects in a "resting state" (eyes covered, ears plugged with cotton). Arterial plasma radioactivities and glucose concentrations are followed for 30–45 min, when PET images of brain radioactivity are obtained. The rate of accumulation of phosphorylated ^{18}FDG in brain is used to calculate $rCMR_{glc}$ (in units of mg 100 g brain^{-1} min^{-1}) by an equation of Sokoloff et al. 1977), modified to take into account dephosphorylation of ^{18}FDG-6-phosphate (Huang et al. 1980). To measure rCBF, the positron emitting water tracer $H_2{}^{15}O$ (radioactive half-life of ^{15}O is 2.03 min) is injected intravenously, and arterial blood radioactivity is sampled throughout a scanning period of 4 min. Local brain concentrations of isotope are measured continuously with PET and are integrated, together with the arterial input function, to generate rCBF images (in ml 100 g brain^{-1} min^{-1}; Frackowiak et al. 1980); Raichle et al. 1984; Alpert et al. 1984).

There are important differences in how $rCMR_{glc}$ and rCBF should be used to examine brain functional activity. As ^{18}FDG disappears slowly from plasma, the $rCMR_{glc}$ method integrates brain activity over a period of at least 30 min, and is inappropriate for studying transient metabolic responses which rapidly habituate (Risberg et al. 1977) during stimulation. It is better used to examine more invariant conditions (e.g., "resting state"). The comparatively long radioactive half-life of ^{18}F (110 min) limits the use of ^{18}FDG to two injections in a single PET study (Chang et al. 1987). Furthermore, because of human radiation exposure limitations, not more than two injections can be administered to the same subject within a three-month period.

On the other hand, human radiation exposure limitations and the short radioactive half-life of ^{15}O permit up to eight rCBF scans (every 12 min) in the same sitting, using $H_2{}^{15}O$. As each scan lasts for 4 min, habituation during activation is less relevant. Control and test-retest scans can be used to examine rCBF during a number of experimental conditions, and difference images of rCBF can be calculated by subtracting one scan from another.

Table 2. $rCMR_{glc}$ ratios between association and primary sensorimotor cortices in AD (data from Haxby et al. 1986)

Metabolic ratio	Control ($N = 30$)	Mild AD ($N = 12$)	Mod AD ($N = 15$)	Severe AD ($N = 8$)
Parietal association sensorimotor	0.93 ± 0.05	0.83 ± 0.09[a]	0.85 ± 0.09[a]	0.72 ± 0.08[a]
Frontal association sensorimotor	0.97 ± 0.07	0.94 ± 0.06	0.92 ± 0.11	0.87 ± 0.18[a]
Lateral temporal association occipital	0.85 ± 0.08	0.76 ± 0.11[a]	0.77 ± 0.11[a]	0.66 ± 0.09[a]

[a] Mean ± SD differs from control by Bonferroni t test ($p < 0.05$).

Correlations Between Resting $rCMR_{glc}$ and Cognitive Test Scores

Selective Vulnerability of Association Neocortex in Relation to Dementia Severity

Differences between mean "resting state" $rCMR_{glc}$ values (mg 100 g^{-1} min^{-1}) in small numbers of mildly demented AD patients and control subjects could not be demonstrated with statistical significance, because of the large standard deviation (SD; 20%–25% of the mean) of these absolute values (Duara et al. 1986). However, SD could be reduced to about 5% of the means by calculating ratios of metabolic rates in association cortices as compared to primary sensory or motor cortices (normalizing for the common influence of global metabolism on both rates). This procedure proved fruitful because it was evident, as illustrated by Fig. 1, that the primary neocortices were much less affected than were the association cortices, basal ganglia and thalamus, even in severely demented patients. As illustrated by Table 2, $rCMR_{glc}$ ratios between the association and primary neocortices demonstrated that the parietal and lateral temporal association cortices were metabolically abnormal in mildly, moderately and severely demented AD patients (Haxby et al. 1986; Rapoport et al. 1986).

Hemispheric Asymmetry of Brain Metabolism and Language/Visuospatial Discrepancy in AD

The heterogeneous patterns and large variances of cognitive test scores and of metabolic rates in AD patients (e.g., Grady et al. 1989) provided a wide range of values which could be examined fruitfully by correlation methods. Two aspects of intersubject metabolic heterogeneity were analyzed by correlation methods: right-left metabolic asymmetries and parietal-prefrontal metabolic ratios.

Right-left metabolic asymmetries were considered in terms of the principle of "cerebral dominance." This principle asserts that one cerebral hemisphere "leads" the other with respect to certain functions. Although cerebral dominance has been called "the most fundamental biological hallmark of human

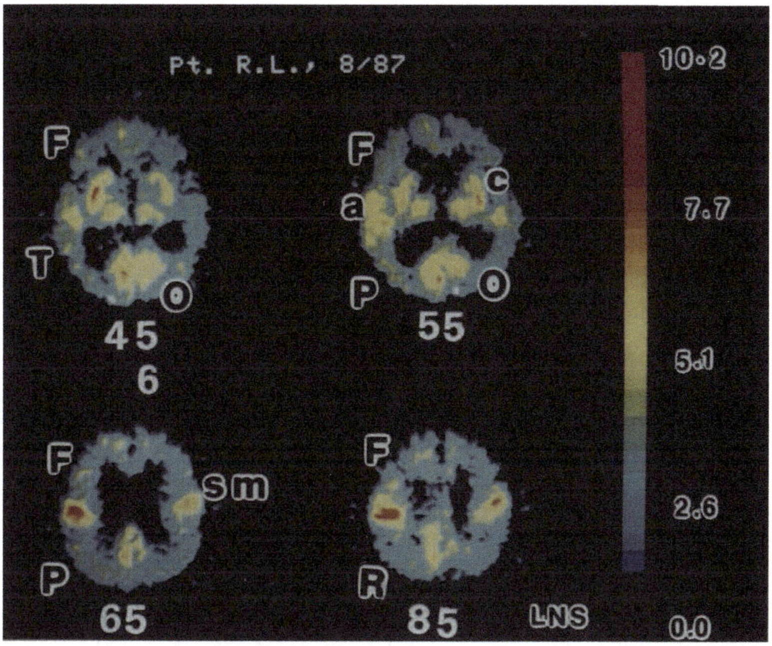

Fig. 1. Positron emission tomography scan from severely demented patient with Alzheimer's disease. $rCMR_{glc}$ derived with Scanditronix tomograph in units of mg 100 g^{-1} min^{-1}. High metabolic rates are retained in the primary sensorimotor cortex (*sm*), primary visual cortex (*o*), primary auditory cortex (*a*) and caudate and lenticular nuclei (*c*), but are reduced in association neocortices. *P*, parietal lobe; *F*, frontal cortex; *T*, temporal cortex; *O*, occipital cortex. Numbers are mm above inferior orbitomeatal line (Laboratory of Neurosciences, National Institute on Aging)

evolution" (Geschwind and Levitsky 1968; Brodal 1981; Passingham 1981), it is found to some extent in the great apes and even in the rhesus monkey (reviewed by Rapoport 1990b). Evidenced by localization of language function in the human left hemisphere, cerebral dominance is genetically determined and occurs in 96% of right-handed and 70% of left-handed or ambidextrous subjects. Consistent with this principle, studies of patients with focal brain damage indicate that syntax comprehension, mental arithmetic and immediate verbal memory are related to left parietal and temporal functions, whereas visuospatial construction is related to right parietal function (Benton 1985; Haxby et al. 1985, 1986).

To see whether marked right-left metabolic asymmetries exist in AD patients, and whether they correlate with appropriate neocortically mediated cognitive deficits, Haxby et al. (1986) used the Syntax Comprehension Test to test left neocortical function, and the Extended Range Drawing Test to examine right neocortical function (cf. Table 1). AD patients were ranked separately on the test scores; the difference between the ranks was calculated as a "syntax/drawing discrepancy." A metabolic asymmetry index (%) was also calculated for each patient as follows:

$$\text{Asymmetry Index (\%)} = \frac{rCMR_{glc}\text{, right} - rCMR_{glc}\text{, left}}{[rCMR_{glc}\text{, right} + rCMR_{glc}\text{, left}]/2} \times 100 \qquad (1)$$

Metabolic asymmetry indices were calculated for mildly and moderately demented AD patients and for healthy controls from $rCMR_{glc}$ data obtained with an ECAT II PET tomograph (ORTEC, Life Sciences, Oak Ridge, TE). As illustrated by the asterisks in Fig. 2, AD patients had significantly greater variances (SD^2) of asymmetry ($p < 0.05$) than controls in the frontal, parietal, and temporal association regions, but not in the sensorimotor or occipital cortices (Haxby et al. 1985), confirming selective vulnerability of these

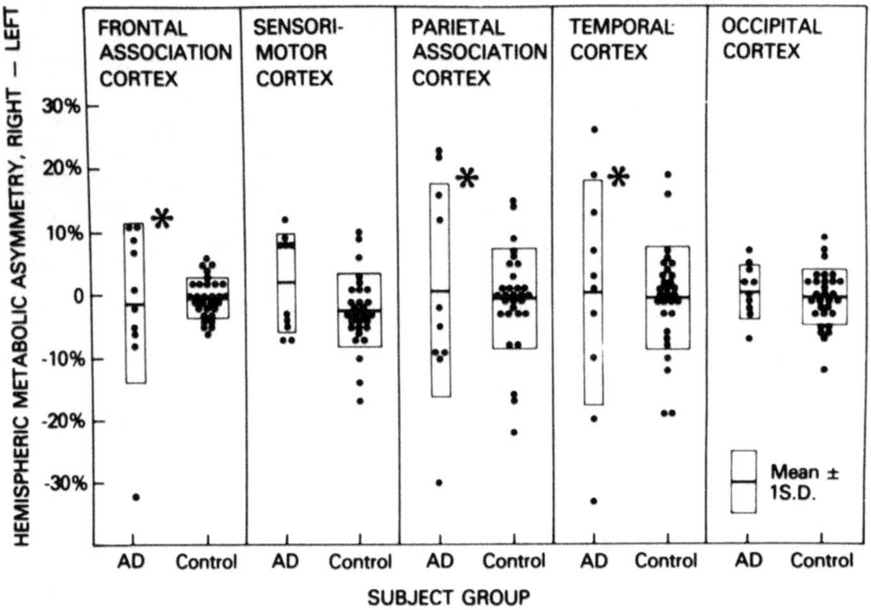

Fig. 2. Metabolic asymmetries (Equation 1, text) in primary and association cortical regions in mildly-moderately demented AD patients and controls. Values a right minus left $rCMR_{glc}$ divided by mean (as %). *, coefficient of variation differs from control value ($p < 0.05$). ECAT data from Haxby et al. (1985)

Table 3. Right-metabolic asymmetry scores in Alzheimer's patients. See Equation 1 for definition (data from Haxby et al. 1986)

Glucose utilization asymmetry index	Control[a] ($N = 29$)	Mild AD ($N = 10$)	Moderate AD ($N = 12$)
Frontal association cortex	0.00 ± 0.03	– 0.01 ± 0.08***	– 0.02 ± 0.12***
Parietal association cortex	0.00 ± 0.06	0.00 ± 0.12*	0.02 ± 0.14***
Lateral temporal association cortex	– 0.01 ± 0.08	– 0.05 ± 0.18**	– 0.04 ± 0.19***

[a] Mean ± SD is given.
Variance is greater than in controls: *$p < 0.05$; **$p < 0.01$; ***$p < 0.001$

association regions. When the patients were grouped according to dementia severity, both the mildly and moderately demented patients were shown to have increased variance of asymmetry in the association neocortices, as compared with controls (Table 3; Haxby et al. 1986). Marked right-left metabolic asymmetries were evident in the PET scans of some patients (Fig. 3).

Consistent with the observation that moderately but not mildly demented AD patients demonstrate mean language and visuospatial deficits (Table 1), the metabolic asymmetry index was correlated significantly and appropriately with syntax/drawing discrepancies in moderately demented patients, but not in the mildly demented patients or controls (Table 4). Correlations were significant for the frontal and parietal association neocortices, but not for the lateral temporal or sensory or motor regions. A lower left-sided $rCMR_{glc}$ corresponded to a

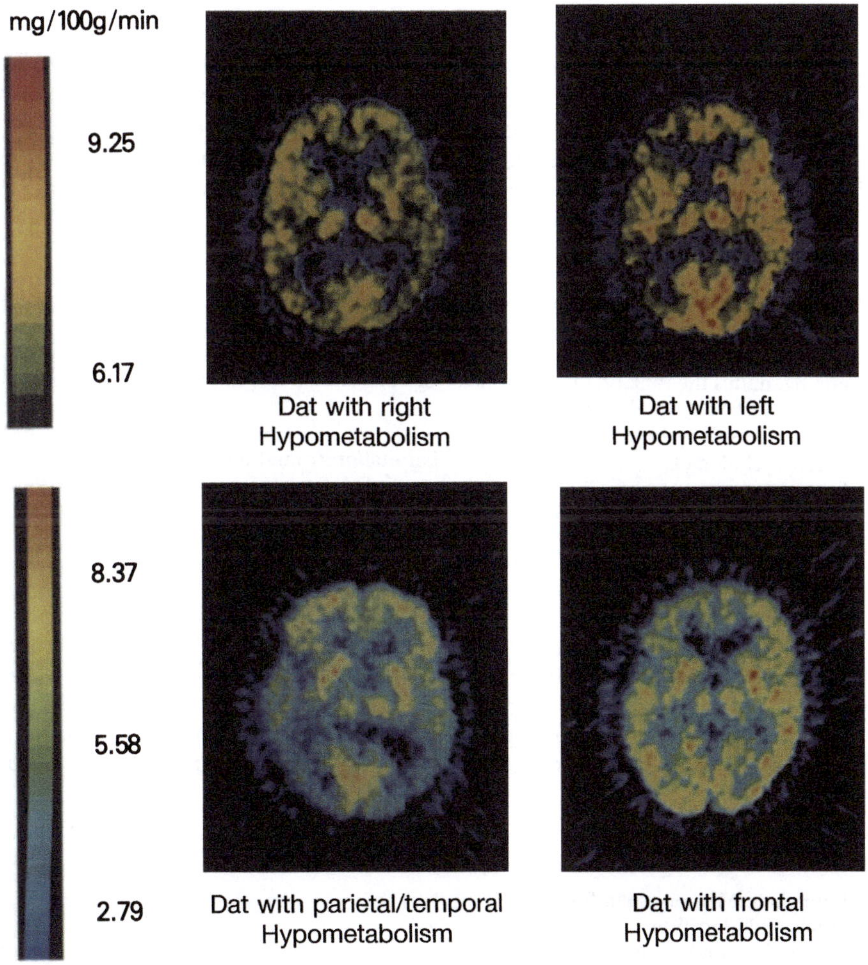

Fig. 3. Examples of extreme metabolic asymmetry (*top*) and of predominant parietal-temporal or frontal hypometabolism in PET scans of different Alzheimer's (DAT) patients. $rCMR_{glc}$ in units of mg 100 g^{-1} min^{-1}. Scanditronix data (Laboratory of Neurosciences, National Institute on Aging)

Table 4. Spearman correlations between right-left metabolic asymmetries (Equation 1) and drawing/syntax comprehension discrepancies in Alzheimer's patients of varying severity of dementia. Positive correlation indicates that better drawing capacity is associated with relatively higher right-sided metabolism (data from Haxby et al. 1986)

Cortical region	Controls (N = 30)	Mild AD (N = 12)	Moderate AD (N = 15)
Frontal association	– 0.30	– 0.01	0.71[a]
Parietal association	– 0.11	– 0.20	0.73[a]
Lateral temporal association	– 0.08	0.01	0.49

[a] Correlation coefficient differs from 0, $p < 0.05$.

worse language score, and a lower right-sided $rCMR_{glc}$ corresponded to a worse visuoconstruction test score. These cross-sectional results suggested that metabolic dysfunction in association regions of AD patients precedes and predicts the deficits in language and visuospatial performance that later appear. Subsequent longitudinal studies confirmed this conclusion (Grady et al. 1986, 1988; Haxby et al. 1990).

Table 5. Spearman correlations between parietal/prefrontal $rCMR_{glc}$ ratios and parietal/prefrontal neuropsychological test score discrepancies in Alzheimer's patients. The discrepancy is the rank on a test score of arithmetic, syntax comprehension, extended range drawing or block tapping, minus the rank on the controlled word association or trailmaking (Time A) test score (data from Haxby et al. 1988)

Neuropsychological discrepancy	Parietal/prefrontal metabolic ratio					
	Control (N = 14–17)		Mild AD (N = 10)		Moderate AD (N = 14)	
	Right	Left	Right	Left	Right	Left
Arithmetic						
Verbal fluency	– 0.02	0.00	– 0.14	0.04	0.63*	0.57*
Attention (Trails A)	– 0.34	– 0.08	– 0.02	– 0.21	0.43	0.46
Verbal comprehension versus						
Verbal fluency	0.00	0.03	– 0.30	– 0.06	0.66*	0.67*
Attention (Trails A)	– 0.12	0.11	0.04	– 0.01	0.51	0.56*
Drawing versus						
Verbal fluency	– 0.02	0.11	– 0.02	0.22	0.66*	0.57
Attention (Trails A)	– 0.21	0.21	0.42	0.32	0.44	0.43
Immediate Memory span for visuospatial location (Block Tapping) versus						
Verbal fluency	0.16	0.05	– 0.25	– 0.01	0.62*	0.54
Attention (Trails A)	0.14	0.32	0.26	0.11	0.73**	0.72**

*$p < 0.05$; **$p < 0.01$

Posterior-Anterior Resting $rCMR_{glc}$ Patterns are Correlated with Discrepancies of Cognitive Functions Mediated by the Frontal and Parietal Lobes

Posterior-anterior metabolic gradients also were found to be correlated significantly with appropriate neuropsychological test discrepancies in AD patients (Haxby et al. 1988). Parietal lobe function was evaluated with the Arithmetic Subtest of the WAIS (Boller and Grafman 1983), the Syntax Comprehension Test, the Extended Range Drawing Test (Benton 1985), and the block tapping span test (DeRenzi and Nichelli 1975; see Table 1). Prefrontal cortex integrity was estimated by the Controlled Word Association (FAS; Jones-Gotman and Milner 1977) and the Trailmaking (Trail A) tests. Cognitive discrepancies representing parietal cortex function as compared with prefrontal cortex function (RANK for performance on a parietal test – RANK for performance on a prefrontal test) were calculated and correlated with parietal association/prefrontal $rCMR_{glc}$ ratios. As illustrated by Table 5, appropriate positive and statistically significant positive correlations were evident between the cognitive discrepancies and the parietal/prefrontal metabolic ratio in moderately but not in mildly demented AD patients.

PET Studies of AD Patients, Related to Brain Network Principles

The above PET studies thus demonstrate that patterns of cerebral metabolic dysfunction in AD patients in a "resting state" are correlated with patterns of specific cognitive deficits. Right-left metabolic asymmetries precede and eventually correlate with discrepancies in visuospatial as compared with language test scores, consistent with the principle of hemispheric dominance. Parietal/prefrontal metabolit ratios correlate with discrepancies in test scores of arithmetic, syntax comprehension, drawing and block tapping on the one hand, and of verbal fluency and attention to simple sets on the other. These observations extend findings on brain-damaged humans and nonhuman primates (Luria 1966, 1973; Jones-Gotman and Milner 1977; Benton 1985; Mesulam 1981), but do not provide details about the neural networks which presumably subserve the cognitive functions. Some of these details can be examined by additional methods which I shall describe below.

Network Hypothesis for the Brain

Network models for the working brain have been proposed by Luria (1973), Mesulam (1981), Goldman-Rakic (1988), Mishkin et al. (1983), Kaas (1989) and Churchland (1986), among others (see above). Indeed, Luria (1973) argued that human mental activites are mediated by "internal" structures – networks or "systems" of interconnected brain regions. Thus, a lesion confined to a single cortical region can cause multiple functional deficits because the region may participate in several partially overlapping networks. Conversely, a cognitive function may be impaired as a consequence of each of several lesions within different components of a critical network.

Correlation Matrix of Resting rCMR$_{glc}$ Data from AD Patients

Based on the network hypothesis, Horwitz et al. (1987) and Horwitz and Rapoport (1988) elaborated a correlation matrix method to examine the covariance of regional cerebral metabolic or flow rates within the brain in subjects in a "resting state" (see above) or performing a task (see below). This method assumes that if two brain regions 1 and 2, are functionally coupled (they belong to a network) so that the activity of one depends on or is related to the activity of the second, a plot of rCMR$_{glc}$ in region 1 against rCMR$_{glc}$ in region 2 (each data point represents a value taken from a single subject) will have a statistically significant correlation coefficient. The size of the correlation coefficient represents the strength of functional coupling between the two regions. Conversely, if two regions are not functionally coupled, the correlation between their rCMR$_{glc}$ values will be small and statistically insignificant.

Figure 4 (left) illustrates a matrix of statistically significant correlation coefficients ($p < 0.025$) obtained from "resting state" rCMR$_{glc}$ data in 21 healthy older subjects, where the correlations were calculated after partialing out the common factor of global CMR$_{glc}$ (Horwitz et al. 1987). The matrix has two characteristics:

1. there are many significant positive correlations between metabolic rates in homologous right and left cortical regions, which appear as diagonals in the right-left lobar submatrices;
2. many correlations occur between frontal and parietal regions, and between occipital and temporal regions, but the frontal-parietal and occipital-temporal "domains" interact minimally.

The AD correlation matrix (Fig. 4, right) differs in several ways from the matrix for elderly healthy controls. It has fewer significant positive partial correlations in the right and left frontal and parietal domains ($p < 0.05$). It also has fewer significant correlations between corresponding regions in the left and right hemispheres, evidenced by fewer correlations in the diagonals of left-right lobar boxes. These differences suggest functional uncoupling between intrahemispheric and interhemispheric neocortical regions in the frontal-parietal domains in AD patients, and are consistent with reduced values of rCMR$_{glc}$ in these association areas.

PET Activation Studies in AD

To further consider network integrity in relation to cognitive function in AD, Grady et al. (1990) administered to seven AD patients and eight age-matched controls a face matching task which had been shown to activate an occipitotemporal association pathway in young healthy subjects (Fig. 5; Haxby et al.

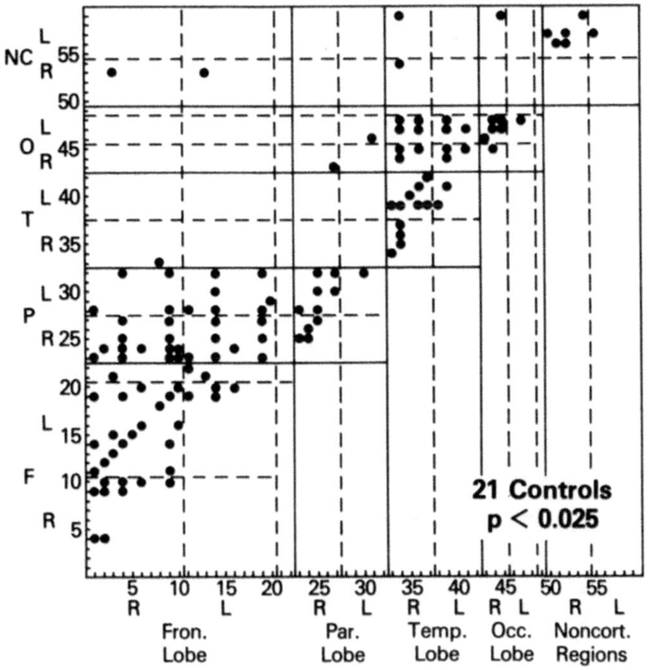

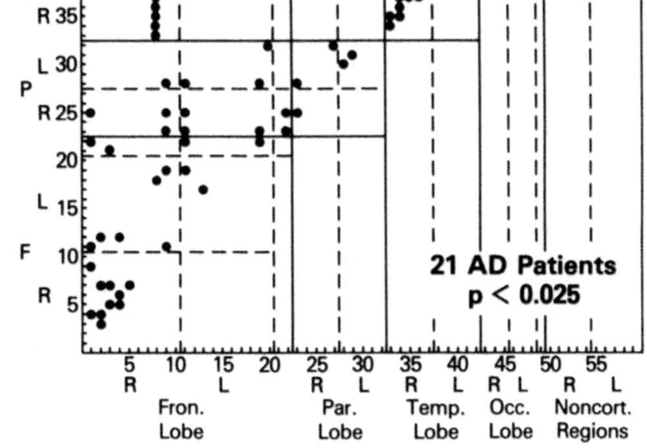

Fig. 4. Matrices of positive partial correlation coefficients between pairs of $rCMR_{glc}$ values for control and Alzheimer's subjects. Correlations at $p < 0.025$ are illustrated. Regions are arranged according to whether they fall in the right (*R*) or left (*L*) frontal (Fron., *F*), parietal (Par., *P*), temporal (Temp., *T*), occipital (Occ., *O*) or noncortical (Noncort., *NC*) domains. See Horwitz et al. (1987) for identities of each numbered region

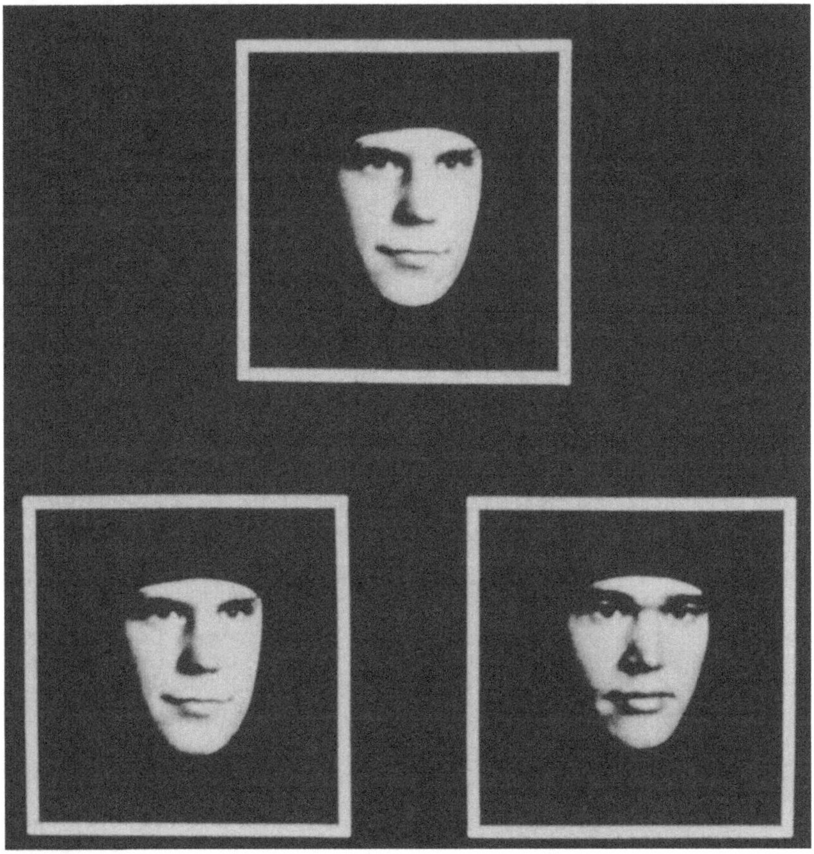

Fig. 5. An example of a face-matching task. Test face is *above*. Choice faces are *below* (from Haxby et al. 1991)

1991). rCBF was measured with $H_2{}^{15}O$ (see above), using a high-resolution PC 1024-7B PET scanner (Scanditronix, Uppsala, Sweden) with an in-plane resolution of 6 mm and an axial resolution of 10 mm. In each subject, rCBF during a control task (which involved pressing a button alternately with the right and left thumbs) was subtracted on a pixel (area element) by pixel basis from rCBF during the face-matching task (which involved pressing the button with the appropriate thumb, after deciding whether the right or left face was to be matched). Flow differences were localized anatomically in a "difference" image. The AD patients and the controls performed the task with accuracies that did not differ significantly (85 ± 8 (SD) % and 92 ± 5%, respectively), but their reaction times (a measure of effort) were more variable than were control times (3323 ± 1464 (SD) ms and 2069 ± 538 ms, respectively).

As illustrated by Fig. 6 (bottom left), scans taken while the AD patients performed the control task frequently demonstrated reduced rCBF in temporoparietal association areas, as compared with scans of normal subjects taken during the control task (Fig. 6, top left), consistent with low "resting state"

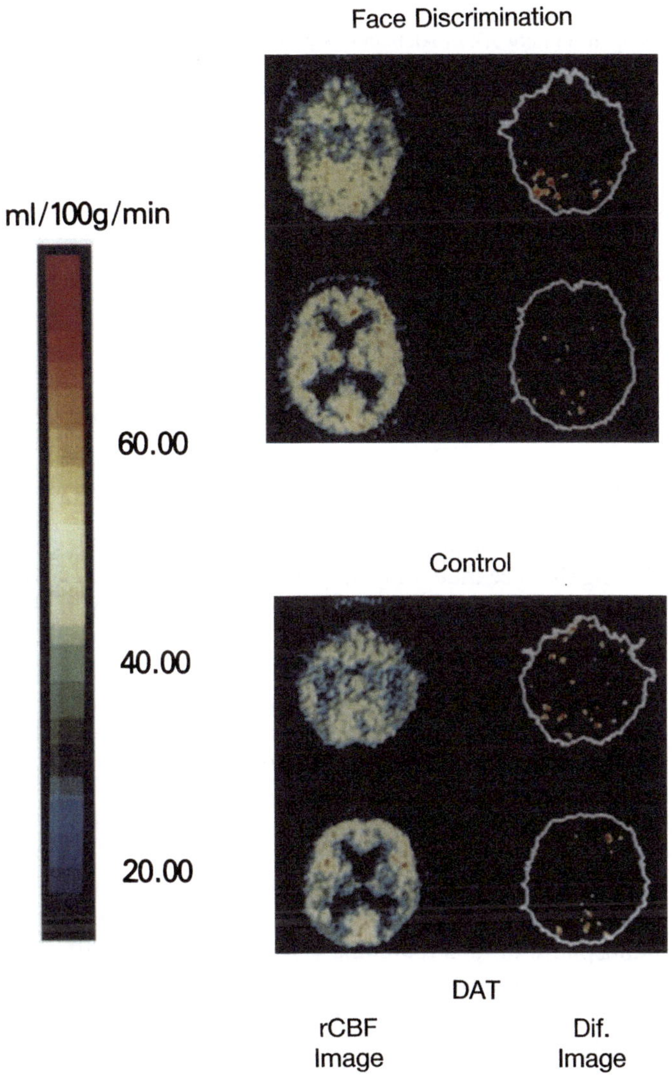

Fig. 6. rCBF activation during a face-matching task in a patient with dementia of the Alzheimer type (DAT) and a control. Scans on *left* show rCBF during the control task (alternate pressing of thumb). Difference images on *right* show rCBF during face-matching (not shown) minus control task rCBF. For DAT and control subjects, lower scans are 50 mm above the inferior orbitomeatal line, top scans are about 25 mm above this line (data are from Grady et al. 1990, and in preparation)

$rCMR_{glc}$ in the association neocortices of AD patients (see above). However, when the patients performed the face matching task, these difference images (Fig. 6, right) showed that initially low-flow association areas were activated bilaterally to the same extent as in healthy age-matched controls (activation measured as a percentage or as absolute change in rCBF). Indeed, even

Table 6. Increments of CBF (as percent of control test baseline) in cortical areas of object recognition network in Alzheimer's patients (AD) and in age-matched controls performing a face-matching task (data from Grady et al. 1990, and in preparation)

Brain region	Age-matched controls (N = 8)	AD (N = 7)
	Percent difference from control task rCBF	
Primary occipital	– 0.8 ± 4.0	2.2 ± 3.0
Occipital association	2.6 ± 2.8	10.6 ± 1.2[a]
Occipitotemporal association	15.4 ± 2.1[a]	17.5 ± 4.1[a]
Superior parietal association	4.6 ± 1.6[a]	4.7 ± 2.3[a]
Frontal association: right	3.4 ± 2.6	11.4 ± 2.6[a]
left	– 2.3 ± 3.0	0.8 ± 1.5

[a] Mean ± SE (bilateral, unless otherwise noted) significantly different from 0, $p < 0.05$.

additional brain areas were activated in the AD patients, perhaps reflecting their greater effort (Grady et al. 1990).

These observations are summarized as mean percent differences in Table 6. During the face matching task, the control subjects activated above control-task rCBF in the occipitotemporal association area by 15.4%, and in the superior parietal association area (usually activated by spatial discrimination tasks; Haxby et al. 1991) by 4.6%. Activation in the AD patients during the task involved, in addition to the occipitotemporal and superior parietal areas, bilateral occipital association and right frontal association (frontal eye field) areas. Thus, the reduced functional activity of visual association regions, evidenced in some patients during the control task, was not accompanied by a reduced capacity of these regions to be activated during object recognition.

Activation of the right frontal eye field area in the AD patients is consistent with its connections with visual association pathways (Pandya and Seltzer 1982; Goldman-Rakic 1988), and with evidence for a dominant role of the right hemisphere in face matching (Damasio et al. 1990). This preference could be demonstrated also by applying a correlation analysis to activated rCBF values during face matching (Horwitz et al. 1990). Despite equivalent bilateral elevations of rCBF in cortical network components of young subjects performing a face matching task, only the rCBF responses in the right hemisphere were found to be correlated significantly with each other. Responses in the left hemisphere were not significantly correlated with each other.

Discussion

Significant correlations between scores on neuropsychological tests in AD patients and "resting state" $rCMR_{glc}$ values, as well as rCBF responses within a cortical network during a face matching task, provide evidence that many aspects of higher cognitive function in humans can be reduced to (mapped to and related to algorithms of) the local structure and integrity of brain networks. These results are consistent with the neurophilosophical contention (Churchland 1986; Changeux, in Changeux and Connes 1990) that critical parameters of mind can be reduced to the structure and function of the brain. They also demonstrate that the relations between mind and brain are disrupted in the course of AD.

Evidence has been presented that visuospatial and linguistic abilities depend on brain networks which are more effective in the right or left hemisphere, respectively, consistent with the principle of "cerebral dominance." The prefrontal association cortex participates in regulating verbal fluency and attention to simple sets, whereas the parietal association cortex is more critical for arithmetic, verbal comprehension, drawing and immediate memory for visuospatial location.

Studies of AD patients and controls performing a face matching task also indicate that a bilateral serial network involving the primary visual (striate) cortex, occipital association and occipitotemporal association areas is critical for carrying out this task. A homologous network subserves object recognition in the rhesus monkey (Mishkin et al. 1983). A correlation analysis of activated rCBF values in subjects performing the face matching task demonstrated more coherent activation in the right than in the left occipitotemporal areas (Horwitz et al. 1990), consistent with a dominant role of the right hemisphere in the face matching process (Damasio et al. 1990).

On the basis of our PET observations during face matching, we recently hypothesized (Rapoport et al. 1991a) that some cognitive deficits in AD are due to failure, initially reversible, of synaptic transmission within the neocortex. Even "resting state" $rCMR_{glc}$ measurements are thought to better represent activity of terminal synapses and their post-synaptic dendritic connections than of neuronal cell bodies or glia (Kadekaro et al. 1985; Sokoloff 1991); suggesting that measured reductions in association cortices of mildly demented AD patients correspond to reduced synaptic transmission. This interpretation is consistent with reduced numbers of intra- and interhemispheric cortical correlations between $rCMR_{glc}$ values in AD patients. Furthermore, normal activation of occipitotemporal association regions in patients performing a face matching task, despite their low "control task" or "idling" rCBF, could reflect reduced synaptic activity at rest but normal activity during face matching. A model for reversible synaptic failure in AD has recently been proposed (Rapoport et al., submitted).

This model is supported by recent neuropathological observations. Synaptic density at lamina III of the frontal association cortex (whose neurons contribute to long cortical-cortical connections, see above) was found to be reduced in biopsy brain tissue from mildly-moderately demented AD patients (Scheff et al. 1990; DeKosky and Scheff 1990). However, the mean length of apposition to postsynaptic membranes was sufficiently enlarged that total synaptic contact area was not abnormal. On the other hand, synaptic density was further reduced in postmortem brains from severely demented AD patients, but apposition length was not further enlarged, so that net contact area was reduced (presumably, synaptic transmission in these cases was irreversibly altered). Deposition of amyloid beta-peptide in the synaptic cleft may contribute to synaptic failure in AD (Koo et al. 1990; Chou et al. 1990).

In summary, whereas knowledge remains limited about the organization and algorithms of brain networks involved in cognitive processing in humans, the combined use of PET and cognitive tests, in normal subjects as well as AD patients, has elucidated and is likely to further elucidate the fundamental neurophilosophical query concerning the relations between mind and brain.

References

Alpert NM, Eriksson L, Chang JY, Bergstrom M, Litton JE, Correia JA, Bohm C, Ackerman RH, Taveras JM (1984) Strategy for the measurement of regional cerebral blood flow using short-lived tracers and emission tomography. J Cereb Blood Flow Metab 4: 28–34

Benton AL (1973) The measurement of aphasic disorders. In: Vilasquez AC (ed) Aspectos patologicas del langage. Centro Neuropsicologico, Lima

Benton A (1985) Visuoperceptual, visuospatial and visuoconstructive disorders. In: Heilman KM, Valenstein E (eds) Clinical neuropsychology, 2nd edn. Oxford University Press, Oxford, pp 186–232

Boller F, Grafman J (1983) Acalculia: historical development and current significance. Brain Cogn 2: 205–223

Brodal A (1981) Neurological anatomy in relation to clinical medicine, 3rd edn. Oxford University Press, Oxford

Chang JV, Duara R, Barker W, Apicella A, Finn R (1987) Two behavioral states studied in a single PET/FDG procedure; theory, method, and preliminary results. J Nucl Med 28: 852–860

Changeux J-P, Connes A (1989) Matière à pensée. Jacob, Paris

Chomsky N (1980) Rules and representations. Behav Brain Sci 3: 1–61

Chou W-G, Zain SB, Rehman S, Tate-Ostroff B, Majocha RE, Benes FM, Marotta CA (1990) Alzheimer cortical neurons containing abundant amyloid mRNA. Relationship to amyloid deposition and senile plaques. J Psychiatr Res 24: 37–50

Churchland PS (1986) Neurophilosophy: toward a unified science of the mind-brain. MIT Press, Cambridge

Cummings JL (1989) Epidemiology of Alzheimer's disease. In: Altman HJ, Altman BN (eds) Alzheimer's and Parkinson's diseases. Recent advances in research and clinical management. Plenum, New York, pp 1–10

Damasio AR, Tranel D, Damasio H (1990) Face agnosia and neural substrates of memory. Annu Rev Neurosci 13: 89–109

DeKosky ST, Scheff SW (1990) Synapse loss in frontal cortex biopsies in Alzheimer's disease: correlation with cognitive severity. Ann Neurol 27: 457–464

DeRenzi E, Nichelli P (1975) Verbal and non-verbal short-term memory impairment following hemispheric damage. Cortex 11: 341–354

Duara R, Grady CL, Haxby JV, Sundaram M, Cutler NR, Heston L, Moore A, Schlageter NL, Larson S, Rapoport SI (1986) Positron emission tomography in Alzheimer's disease. Neurology 36: 879–887

Folstein MF, Folstein SE, McHugh PR (1975) "Mini-mental state." A practical method for grading the cognitive state of patients for the clinician. J Psychiatr. Res 12: 189–198

Frackowiak RSJ, Lenzi GL, Jones T, Heather JD (1980) Quantitative measurement of regional cerebral blood flow and oxygen metabolism in man using ^{15}O and positron emission tomography: theory, procedure and normal values. J Comput Assist Tomogr 4: 727–736

Gallup GG Jr (1977) Self-recognition in primates. A comparative approach to the bidirectioinal properties of consciousness. Am Psychol 32: 329–338

Geschwind N, Levitsky W (1968) Human brain: left-right asymmetries in temporal speech region. Science 161: 186–187

Golden CJ (1978) The Stroop color and word test: a manual for clinical and experimental uses. Stoelting, Chicago

Goldman-Rakic PS (1988) Changing concepts of cortical connectivity: parallel distributed cortical networks. In: Rakic P, Singer W (eds) Neurobiology fo neocortex. Wiley, New York, pp 177–202

Gove PB (1966) Websters third new international dictionary of the English language, unabridged. Miriam, Springfield, p 1698

Grady CL, Haxby JV, Schlageter NL, Berg G, Rapoport SI (1986) Stability of metabolic and neuropsychological asymmetries in dementia of the Alzheimer type. Neurology 36: 1390–1392

Grady CL, Haxby JV, Horwitz B, Sundaram M, Berg G, Schapiro M, Friedland RP, Rapoport SI (1988) A longitudinal study of the early neuropsychological and cerebral metabolic changes in dementia of the Alzheimer type. J Clin Exp Neuropsychol 10: 576–596

Grady CL, Haxby J, Schapiro MB, Kumar A, Friedland RP, Rapoport SI (1989) Heterogeneity in dementia of the Alzheimer type (DAT): subgroups identified from cerebral metabolic patterns using positron emission tomography (PET). Neurology 39 [Suppl 1]: 167–168

Grady CL, Haxby J, Horwitz B, Schapiro M, Carson R, Herscovitch P, Rapoport S (1990) Activation of regional cerebral blood flow (rCBF) in extra-striate cortex curing a face matching task in patients with dementia of the Alzheimer type (DAT). Soc Neurosci Abstr 16: 149

Grundke-Iqbal I, Iqbal K, Tung Y-C, Quinlan M, Wisniewski HM, Binder LI (1986) Abnormal phosphorylation of the microtubule-associated protein τ (tau) in Alzheimer cytoskeletal pathology. Proc Natl Acad Sci USA 83: 4913–4917

Haxby JV, Duara R, Grady CL, Cutler NR, Rapoport SI (1985) Relations between neuropsychological and cerebral metabolic asymmetries in early Alzheimer's disease. J Cereb Blood Flow Metab 5: 193–200

Haxby JV, Grady CL, Duara R, Schlageter N, Berg G, Rapoport SI (1986) Neocortical metabolic abnormalities precede non-memory cognitive defects in early Alzheimer's type dementia. Arch Neurol 43: 882–885

Haxby JV, Grady CL, Koss E, Horwitz B, Schapiro M, Friedland RP, Rapoport SI (1988) Heterogeneous anterior-posterior metabolic patterns in Alzheimer's type dementia. Neurology 38: 1853–1863

Haxby JV, Grady CL, Koss E, Horwitz B, Schapiro MB, Katz D, Friedland RP, Rapoport SI (1990) Longitudinal study of cerebral metabolic asymmetries and associated neuropsychological deficits in early dementia of the Alzheimer type. Arch Neurol 47: 753–760

Haxby JV, Grady CL, Horwitz B, Ungerleider LG, Mishkin M, Carson RE, Herscovitch P, Schapiro MB, Rapoport SI (1991) Dissociation of spatial and object visual processing pathways in human extrastriate cortex. Proc Natl Acad Sci USA 88: 1621–1625

Heston LL, Mastri AR, Anderson VE, White J (1981) Dementia of the Alzheimer type. Clinical genetics, natural history, and associated conditions. Arch Gen Psychiatry 38: 1085–1090

Horwitz B (1990) Quantification and analysis of positron emission tomography metabolic data. In: Duara R (ed) Positron emission tomographfy in dementia. Wiley-Liss, New York, pp 2–70 (Frontiers of clinical neuroscience, vol 10)

Horwitz B, Rapoport SI (1988) Partial correlation coefficients approximate the real intrasubject correlation pattern in the analysis of interregional relations of cerebral metabolic activity. J Nucl Med 29: 392–399

Horwitz B, Grady CL, Schlageter NL, Duara R, Rapoport SI (1987) Intercorrelations of regional cerebral glucose metabolic rates in Alzheimer's disease. Brain Res 407: 294–306

Horwitz B, Haxby J, Grady C, Schapiro M, Carson R, Herscovitch P, Ungerleider L, Mishkin M, Rapoport SI (1990) Interregional correlations of regional cerebral blood flow (rCBF) among posterior cortical brain regions during object and spatial visual processing. Soc Neurosci Abstr 16: 28

Huang S-C, Phelps ME, Hoffman EJ, Sideris K, Selin CJ, Kuhl DE (1980) Non-invasive determination of local cerebral metabolic rate of glucose in man. Am J Physiol 238: E69–E82

Jones-Gotman M, Milner B (1977) Design fluency: the invention of nonsense drawings after focal cortical lesions. Neuropsychologia 15: 653–674

Kaas JH (1989) Why does the brain have so many visual areas? J Cogn Neurosci 1: 121–135
Kadekaro M, Crane AM, Sokoloff L (1985) Differential effects of electrical stimulation of sciatic nerve on metabolic activity in spinal cord and dorsal root ganglion in the rat. Proc Natl Acad Sci USA 82: 6010–6013
Katzman R (1976) The prevalence and malignancy of Alzheimer's disease: a major killer. Arch Neurol 33: 217–218
Koo EH, Sisodia SS, Archer DR, Martin LJ, Weidemann A, Beyreuther K, Fischer P, Masters CL, Price DL (1990) Precursor of amyloid protein in Alzheimer disease undergoes fast anterograde axonal transport. Proc Natl Acad Sci USA 87: 1561–1565
Lewis DA, Campbell MJ, Terry RD, Morrison JH (1987) Laminar and regional distributions of neurofibrillary tangles and neuritic plaques in Alzheimer's disease: a quantitative study of visual and auditory cortices. J Neurosci 7: 1799–1808
Luria AR (1966) Higher cortical functions in man. Basic, New York
Luria AR (1973) The working brain. An introduction to neuropsychology. Basic, New York
Mattis S (1976) Mental status examination for organic mental syndrome in the elderly patient. In: Bellack L, Karasu TB (eds) Geriatric psychiatry: a handbook for psychiatrists and primary care physicians. Grune and Stratton, New York, pp 77–121
McKhann G, Drachman D, Folstein M, Katzman R, Price D, Stadlan EM (1984) Clinical diagnosis of Alzheimer's disease: report of the NINCDS-ADRDA Work Group under the auspices of Department of Health and Human Services Task Force on Alzheimer's disease. Neurology 34: 939–944
Menzel EW (1973) Chimpanzee spatial memory organization. Science 182: 943–945
Mesulam MM (1981) A cortical network for directed attention and unilateal neglect. Ann Neurol 10: 309–325
Milner B (1971) Interhemispheric differences in the localization of psychological process in man. Br Med Bull 27: 272–277
Mishkin M, Ungerleider LG, Macko KA (1983) Object vision and spatial vision: two cortical pathways. Trends Neurosci 6: 414–417
Palmert MR, Golde TE, Cohen ML, Kovacs DM, Tanzi RE, Gusella JF, Usiak MF, Younkin LH, Younkin SG (1988) Amyloid protein precursor messenger RNAs: differential expression in Alzheimer's disease. Science 241: 1080–1084
Pandya DN, Seltzer B (1982) Association areas of the cerebral cortex. Trends Neurosci 5: 386–390
Passingham RE (1981) Broca's area and the origins of human vocal skill. Philos Trans R Soc Lond [Biol] 292: 167–175
Penrose R (1989) The emperor's new mind: concerning computers, minds, and the laws of physics. Oxford University Press, Oxford
Porteus SD (1965) Porteus maze test: fifty years' application. Pacific, Palo Alto
Premack AJ, Premack D (1972) Teaching language to an ape. Sci Am 227: 92–99
Raichle ME, Martin WR, Herscovitch P, Mintun MA, Markham J (1984) Brain blood flow measured with intravenous H2(15)O. II. Implementation and validation. J Nucl Med 24: 790–798
Rapaport SI (1988) Brain evolution and Alzheimer's disease. Rev Neurol (Paris) 144: 79–90
Rapoport SI (1989) Hypothesis: Alzheimer's disease is a phylogenic disease. Med Hypotheses 29: 147–150
Rapoport SI (1990a) Topography of Alzheimer's disease: involvement of association neocortices and connected regions; pathological, metabolic and cognitive correlations; relation to evolution. In: Rapoport SI, Petit H, Leys D, Christen Y (eds) Imaging, cerebral topography and Alzheimer's disease. Research and perspectives in Alzheimer's disease. Springer, Berlin Heidelberg New York, pp 1–17
Rapoport SI (1990b) Integrated phylogeny of the primate brain, with special reference to humans and their diseases. Brain Res 15: 267–294
Rapoport SI, Horwitz B, Haxby JV, Grady CL (1986) Alzheimer's disease: metabolic uncoupling of associative brain regions. Can J Neurol Sci 13: 540–545

Rapoport SI, Horwitz B, Grady CL (1991a) Hypothesis: resting and activated PET scans of Alzheimer patients suggest reversible failure of synaptic transmission (Abstr). J Cereb Blood Flow Metab (in press)
Rapoport SI, Pettigrew KD, Schapiro MB (1991b) Discordance and concordance of dementia of the Alzheimer type (DAT) in monozygotic twins indicate heritable and sporadic forms of Alzheimer's disease. Neurology 41: 1549–1553
Reitan RM (1958) Validity of the trail making test as an indicator of organic brain damage. Percept Mot Skills 8: 271–276
Roofe PG, Matzke HA (1968) Introduction to the study of evolution: its relationship to neuropathology. In: Minkler J (ed) Pathology of the nervous system, vol 1. Blakiston, New York, pp 14–22
Risberg J, Maximilian VA, Prohovnik I (1977) Changes of cortical activity patterns during habituation to a reasoning test. Neuropsychologica 15: 793–798
Roses AD, Yamoaka L, Gaskell PC, Bebout JL, Alberts MJ, Haynes CS, Hung W-Y, Welch K, Earl N, Clark C, Pericak-Vance MA (1990) Linkage analysis of late onset familial Alzheimer's disease. Neurobiol Aging 11: 319
Roy CS, Sherrington CS (1890) On the regulation of the blood supply to the brain. J Physiol (Lond) 11: 85–108
Russell EW (1975) A multiple scoring method for the assessment of complex memory functions. J Consult Clin Psychol 43: 800–809
Schapiro MB, Haxby JV, Grady CL, Rapoport SI (1990) Topographical comparison of lesions in trisomy 21 and Alzheimer's disease: a study with PET, anatomical and neuropathological investigations. In: Rapoport SI, Petit H, Leys D, Chrisen Y (eds) Imaging, cerebral topography and Alzheimer's disease. Research and Perspectives in Alzheimer's disease. Springer, Berlin Heidelberg New York, pp 97–108
Scheff SW, DeKosky ST, Price DA (1990) Quantitative assessment of cortical synaptic density in Alzheimer's disease. Neurobiol Aging 11: 29–37
Selkoe DJ, Bell DS, Podlisny MB, Price DL, Cork LC (1987) Conservation of brain amyloid proteins in aged mammals and humans with Alzheimer's disease. Science 235: 873–877
Sokoloff L (1991) Relationship between functional activity and energy metabolism in the nervous system: whether, where and why. In: Ingvar DH, Lassen NA, Raichle ME, Friberg L (eds) Brain work II. Munksgaard, Copenhagen (Alfred Benzon symposium II) (in press)
Sokoloff L, Reivich M, Kennedy C, Des Rosiers MH, Patlak CS, Pettigrew KD, Sakurada O, Shinohara M (1977) The [14C]deoxyglucose methods for the measurements of local cerebral glucose utilization: theory, procedure, and normal values in the conscious and anesthetized albino rat. J Neurochem 28: 897–916
St George-Hyslop PH, Tanzi RE, Polinsky RJ, Haines JL, Nee L, Watkins PC, Myers RH, Feldman RG, Pollen D, Drachman D, Growdon J, Bruni A, Foncin J-F, Salmon D, Frommelt P, Amaducci L, Sorbi S, Piancentini S, Stewart GD, Hobbs WJ, Conneally PM, Gusella JF (1987) The genetic defect causing familial Alzheimer's disease maps on chromosome 21. Science 235: 885–890
Taylor J (ed) (1931) Selected writings of John Hughlings Jackson. Evolution and dissolution of the nervous system, vol 2. Basic, Newe York, pp 3–118
Tellegen A, Briggs PF (1967) Old wine in new skins: grouping Wechsler subtests into new scales. J Consult Psychol 31: 499–506
Terry RD, Wisniewski HM (1972) Ultrastructure of senile dementia and of experimental analogs. In: Gaitz CM (ed) Aging and the brain. Plenum, New York, pp 89–116 (Advances in Behavioral Biology, vol 3)
Wechsler DA (1945) A standardized memory scale for clinical use. J Psychol 19: 87–95
Wechsler DA (1955) Manual for the Wechsler adult intelligence scale. Psychological Corporation, New York
Wise SP, Jones EP (1977) Cells of origin and terminal distribution of descending projections of the rat somatic sensory cortex. J Comp Neurol 175: 129–158
Wisniewski KE, Wisniewski HM, Wen GY (1985) Occurrence of neuropathological changes and dementia of Alzheimer's disease in Down's syndrome. Ann Neurol 17: 278–282

Memory, Language and Decision-Making: Contributions from the Lesion Method in Humans*

H. Damasio and *A. R. Damasio*

Summary

The advent of fine structural neuroimaging techniques such as x-ray computerized tomography (CT) and magnetic resonance (MR) now permits detailed neuroanatomical studies of humans *in vivo*. This development has made the lesion method viable in humans with neurological lesions. Combined with new techniques for cognitive experimentation and theoretical advances on the conceptualization of the nervous system, the lesion approach has led to a new understanding of the neural systems underlying learning and memory, language, and the processes of decision-making and planning.

Introduction

The modern practice of the lesion method requires four ingredients:
1. a theory about the architecture and operation of the normal brain;
2. a hypothesis about the function of a subsystem of that normal brain;
3. an experimental task related to the hypothetical function of the subsystem; and
4. a lesion in a component of the subsystem. In essence a lesion is a probe to test the validity of a hypothesis.

The lesion method is one of the oldest in the history of neuroscience and has consistently contributed major advances to the understanding of the primate brain. For a time, especially during the middle part of the century, the approach lost favor. This was perhaps because other approaches yielded results faster, or because the lesion method was so often associated with traditional, and by then not acceptable, accounts of brain function. More recently, however, both in experimental animals and in humans, the lesion method has been inspired by modern theory (and the results contributed to theoretical advancement).

* Supported by NINCDS Grant PO1 NS19632 and NIH Grant RR59. Part of this text was published in *Seminars in the Neurosciences*, Vol. 2, 1990, under the title "The Neural Basis of Memory, Language and Behavioral Guidance: Advances with the Lesion Method in Humans

Furthermore, the advent of imaging methods such as x-ray computerized tomography (CT) and magnetic resonance (MR) has led to fine neuroanatomical detailing of the human brain in vivo. In turn, that has made the lesion approach experimentally viable in neurological patients, and finally permitted a clear distinction between interesting clinical studies of single neurological patients, on the one hand, and fundamental, systematic neuropsychological studies on the other.

In this review we outline some developments in the fields of memory, language, and behavioral guidance that have come about as a result of the lesion method in humans. A review of progress with this same approach in animals is outside the scope of this chapter but the reader is referred to comprehensive surveys by Desimone and Ungerleider (1989) and Mishkin and Appenzeller (1987) which cover developments in vision and memory, respectively.

Memory

The understanding of the neural basis of memory was altered three decades ago by a lesion study in the patient HM. HM had a bilateral resection of his medial temporal region performed to relieve him of intractable seizures (Scoville and Milner (1957). It was predicted that the operation would reduce the seizures, and that was indeed the case, but it was not known at the time that a bilateral resection of this region would cause a severe amnesia. The study suggested (and later cases confirmed) that bilateral damage to the hippocampus and nearby structures precluded the learning of any new information at both episodic and generic levels. [Episodic information refers to unique objects and events; generic (also known as semantic) information denotes objects and events as members of a category]. Information that had been learned before the surgical intervention, however, could still be retrieved in recall and used for recognizing unique persons or objects in general, i.e., to recognize at both episodic and generic levels.

A large number of subsequent studies performed in patient HM and in other patients with disorders of learning and memory has built on those seminal findings, especially over the past decade. The fundamental advances are as follows:

1. The type of information that patient HM is unable to learn consists of both entities and events as members of a category and as unique exemplars. In other words, it ranges from objects to persons, in isolation or when they interact in a given scene. Such information has been called *declarative* and requires, when we recall it, the activation of an internal representation of sensory data describing the particular object or person. Another type of information, however, consists of perceptual and motor strategies and enables us to perform actions with a particular pattern, e.g., swimming, riding a bicycle, typing, reading a sentence given its mirror image, etc. It does not deal with facts, as declarative memory does, but rather with skills, and it has been called *procedural.* Rather than an internal representation, the retrieval of procedural information requires a motor output.

It was thus important to discover that patient HM (and others who have also lost the hippocampal system bilaterally) did not have any difficulty learning new perceptuomotor skills, a clear indication that the hippocampal system is not required for such learning (Milner et al. 1968; Cohen and Squire 1980; Damasio et al. 1987). This powerful dissociation can even be seen in patients with Alzheimer's disease, in whom the memory defect is largely due to bilateral dysfunction in the hippocampal system (Eslinger and Damasio 1986). It is important to note that, while all of these amnesic patients learn a skill and can perform it in a manner comparable to controls, they never recall the circumstances in which their learning took place. In other words, they do not remember ever seeing the experimental apparatus, nor do they recall being given instructions for the task. Naturally, the experimental setting and the instructions are factual, declarative types of information, which require the activation of a conscious internal representation. These patients seem incapable of achieving such representations or, at least, of making them conscious.

2. Patient HM was able to recall generic as well as episodic information from his past. For instance, he could recognize places, persons and events learned before the age of 16 (Corkin 1984). It is thus of great importance to know that another well-known amnesic patient, Boswell, can recall little episodic information from his entire past, and that he cannot place in appropriate temporal frame the few items that he does recall. He is unable to recognize family or friends, places or objects from his past and he cannot narrate any specific episode, however important, from the several decades that preceded the onset of his lesion. He is, in short, unable to reconstitute his autobiography. His severe retrograde amnesia spans all the way from age 44, when he became amnesic, back to the very early years of his life (Damasio et al. 1985, 1987). What is the anatomical basis for this remarkably different profile of amnesia? The answer lies in the extensive damage to higher order association cortices located in both anterior temporal regions, which are not part of the hippocampal system itself. While patient HM has suffered damage to hippocampus, and to part of the parahippocampal gyrus and amygdala, the damage in patient Boswell extends beyond those structures to encompass area 38 (the temporal pole), areas 21 and 20 and part of area 37 (a territory that together constitutes the so-called "inferotemporal" region), and part of area 22 (Fig. 1). The findings in Boswell establish that those cortices support access to the entire range of episodic knowledge which has been learned with the assistance of the hippocampal system. Such knowledge is clearly *not* inscribed within the hippocampal system.
3. As noted above, the damage in patient HM covered three important structures: the hippocampal formation itself, some nuclei of the amygdala, and the parahippocampal gyrus, especially in its anterior portion (known as entorhinal cortex or area 28). Given the interlocking circuitry of these three regions (Van Hoesen and Pandya 1975a, b; Van Hoesen et al. 1975; Rosene and Van Hoesen 1977; Van Hoesen 1982; Amaral 1987), it is likely that HM's learning defect is the result of combined damage, but the possibility that each of these three key structures could cause a learning defect *per se* remains

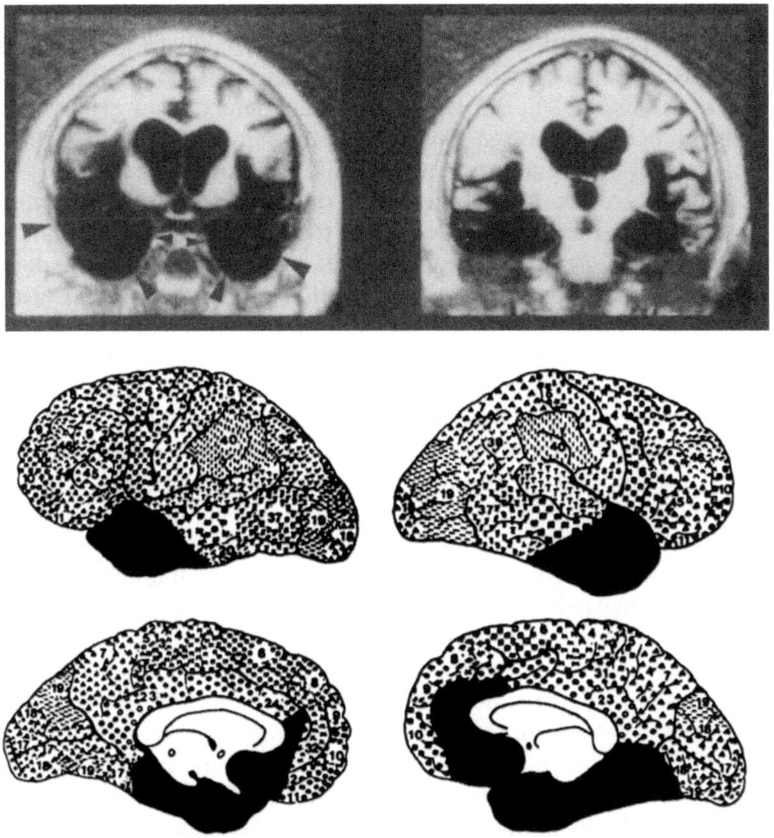

Fig. 1. Magnetic resonance scan (T_1-weighted pulse sequence) of patient Boswell's brain. The *top left section* reveals the complete destruction of the anterior temporal cortices, both in their medial and lateral aspects *(black areas highlighted by arrows)*. The *top right section* reveals the preservation of middle and posterior temporal cortices. The *bottom diagram* reveals the full extent of the damage marked in black. Both left and right hippocampal systems are destroyed (the hippocampal formation, the entorhinal cortex, and amygdala), along with cytoarchitectonic fields 38, 20 and 21, 35 and 36, and part of medial 37. The early sensory cortices for the visual, auditory, and somatosensory modalities are intact and so are several intermediate cortices in areas 39, 40, and part of 37. Most of the frontal association cortices are preserved

open. Some important new information in this regard is now available. Firstly, we know that damage confined to a sector of the hippocampus will still impair learning. Such a defect is consistently seen in patients who sustain cerebral anoxia and ischemia, and suffer cell loss in the CA1 territory of both hippocampal formations. A detailed neuropathological confirmation of the circumscribed nature of this lesion is now available (Zola-Morgan et al. 1986). Secondly, we know that damage to the entorhinal cortex in Alzheimer's disease occurs massively and early on in the disease and is the probable cause of the early manifestation of memory decline in the affected patients. The damage involves the layers that provide high order cortical input to hippocampus (layer II) and output from hippocampus to cortex (layer IV), thus disrupting the feedforward/feedback loop on which cortical-hippocampal interactions are based (Van Hoesen and Damasio 1987). This sort of damage and the ensuing learning defect occur even when damage to the hippocampal formation itself is minimal. In sum, each of these structures makes an independent contribution to learning mechanisms.

4. Damage confined to structures of the basal forebrain in humans causes a defect of learning and retrieval that is distinguishable from the defect caused by damage in temporal structures (Damasio et al. 1985). The defect is caused by infarctions in the territory of the anterior communicating artery or anterior cerebral artery, generally in the setting of ruptured congenital aneurysms in those arteries. The basal forebrain area is critical for memory on a variety of counts. Firstly, it contains numerous nuclei that provide most of the intrinsic acetylcholine innervation of the cerebral cortex, e.g., the nucleus basalis of Meynert within the substantia innominata, the nucleus accumbens, and the septal nuclei. Secondly, the region is traversed by the axons of non-specific neurotransmitter systems in their course from brain stem nuclei to the cerebral cortices. Thirdly, this region is closely interconnected with the hippocampal system in the medial temporal region. Patients with damage in basal forebrain tend to recall poorly material learned in both the anterograde (after lesion) and the retrograde (before lesion) temporal compartments. A critical feature that distinguishes these cases is the remarkable improvement of recognition and recall performances when the subject is helped with cueing. The discovery of an association between basal forebrain damage and amnesia enlarges our knowledge about the role of subcortical structures in memory, which has previously implicated the medial nuclei of the thalamus and the hypothalamus (mammillary bodies) in studies of Wernicke Korsakoff syndrome (this is a syndrome that develops in thiamine-deficient alcoholics and causes a severe amnesia; Butters and Cermak 1980; Victor et al. 1971; Mair et al. 1979).
5. The study of defects in face recognition has offered new insights into the processes of learning and memory. The topic is of special importance because, in most instances, patients unable to recognize faces are also unable to learn new faces (Damasio et al. 1982a, 1990b). In patients in whom the defect is due to damage in bilateral visual association cortices that are far removed from both anterior temporal cortices and the hippocampal system, this is an indication that the neural systems that support perception are also critical for the recording of new material regarding those faces, i.e., the representation of the visual characteristics of new faces and the link between such representations and information that is uniquely pertinent to those faces. The study of face processing has yielded orther important clues. For instance, face recognition and new face learning can be disrupted not only by damage within posterior visual association cortices but also by damage within anterior temporal cortices. This is a strong indication that the neural systems necessary for the learning and eventual recognition of faces require multiple components that encompass early visual association cortices, intermediate visual association cortices, higher-order cortices, and the hippocampal

system itself. Finally, the study of defects in face processing has shown that patients who do not recognice any familiar face consciously can still perform nonconscious, covert recognition. The strongest evidence in this area comes from discriminatory skin conductance responses generated by face agnosic patients in a passive paradigm developed by Tranel and Damasio (1985, 1988). The patients were merely asked to view familiar or non-familiar faces without the need to make a verbal response and without a supplementary name cue.

6. Perhaps the most exciting new contributions of the lesion method to memory come from the discovery of category-related defects in recognition associated with damage to the temporal cortices (Fig. 2; Damasio 1990; Damasio et al. 1990a). The recent evidence offers neuroanatomical correlates for the dissociations in retrieval of generic knowledge that have been described by investigators in other laboratories (Warrington and Shallice 1984; McCarthy and Warrington 1988) and in ours. In brief, it appears that access to knowledge about entities that are learned through the visual modality alone and are visually "ambiguous" [i. e., that share physical structures with several other different entities; typical examples are animals of the wolf family, i. e., fox, raccoon, coyote, whose physical traits strongly resemble each other] depends on the inferotemporal (IT) region of both hemispheres (as noted above, the IT territory includes areas 21, 20 and 37). Knowledge of visual entities that have lesser ambiguity (an elephant or a giraffe are good examples) does not depend on this region, nor does knowledge of entities that were learned through *both* the visual and somatosensory modalities (manipulable tools and utensils are the prime examples). The evidence for this finding comes from studies in which patients with bilateral or unilateral damage to IT had profound defects for the recognition of a large number of animals but were largely intact in their recognition of tools and utensils. These studies reveal that access to previously acquired generic (semantic) knowledge can be disrupted by lesions in specific neural subsystems, and that

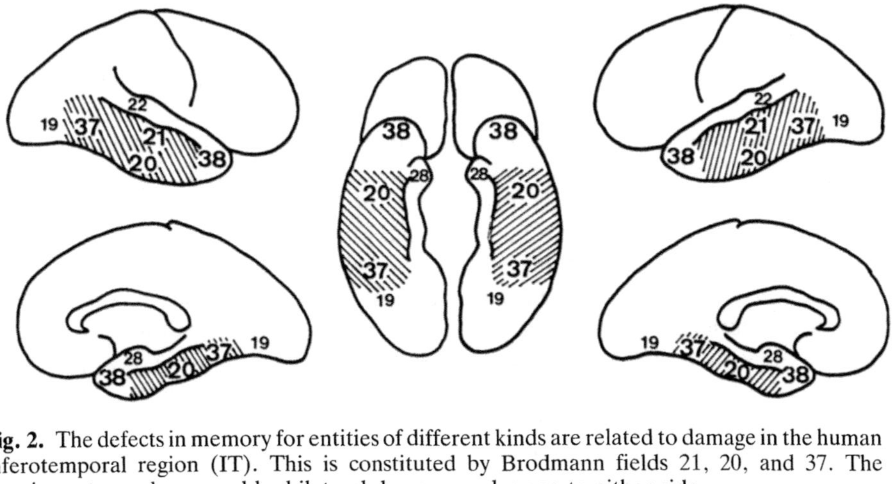

Fig. 2. The defects in memory for entities of different kinds are related to damage in the human inferotemporal region (IT). This is constituted by Brodmann fields 21, 20, and 37. The impairments can be caused by bilateral damage or damage to either side

the defect is not equal across all types of knowledge. This provides indirect evidence that different types of knowledge are laid down, in a distributed manner, in different neural systems. The consistent choice of those systems for certain types of knowledge probably depends on constraints dictated by the entities themselves and by the brain's design.

The findings reviewed above contribute to a far more detailed picture of the neural systems underlying learning and memory than was available before. The role of the hippocampus in learning has been reconfirmed, but it appears that the hippocampus is not essential for recall and recognition. The highly distributed neural inscriptions on the basis of which generic and episodic knowledge are recorded are located along a complex hierarchy of cortices placed all the way from the vicinity of the primary sensory portals to the higher-order fields progressively closer to the hippocampal system (Damasio 1989a, b). A variety of subcortical nuclei (the neurotransmitter nuclei and brain stem, the basal forebrain, and the medical thalamus and mammillary bodies) appear to assist the above-mentioned cortices and the hippocampal system during learning and retrieval. Finally, it is clear that the learning and maintenance of perceptual motor skills do not depend on the hippocampal system and must depend instead on subcortical structures in cerebellum and basal ganglia, together with somatosensory and motor cortices in parietal and frontal regions.

Language

During the past decade the lesion method has brought significant progress to the study of language. The new studies have demonstrated that the neural networks on which language depends include structures, both cortical and subcortical, that had not been a part of the traditional map of language-related areas, which is largely based on aphasia studies. The traditional map includes Wernicke's area (which, in its narrow definition, is made up by the posterior sector of Brodmann's area 22 in the left hemisphere), Broca's area (areas 44 and 45 in the left hemisphere), and the left supramarginal and angular gyri (respectively, areas 40 and 39). The language map that is surfacing from recent lesion studies is far richer.

The most interesting development comes from the discovery that the anterior sector of the left temporal cortices (areas 38, 21, 20 and 37; Fig. 3) is dedicated to language. The systems in this territory are not concerned with grammatical or phonemic/phonetic aspects of language and probably constitute the neural basis for the reference lexicon (the collection of words that denotes concrete entities and actions). A recent study based on lesions caused by herpes encephalitis and surgical ablations (Damasio et al. 1990a) establishes the following:

1. Damage to the left anterotemporal sector including the temporal pole (area 38) and the anterior part of the inferotemporal region (areas 21, 20, 37), causes a severe defect for naming of concrete entities. Patients with such lesions cannot access the proper names that go with unique entities, nor the common nouns that go with varied entities, natural or man-made, at the

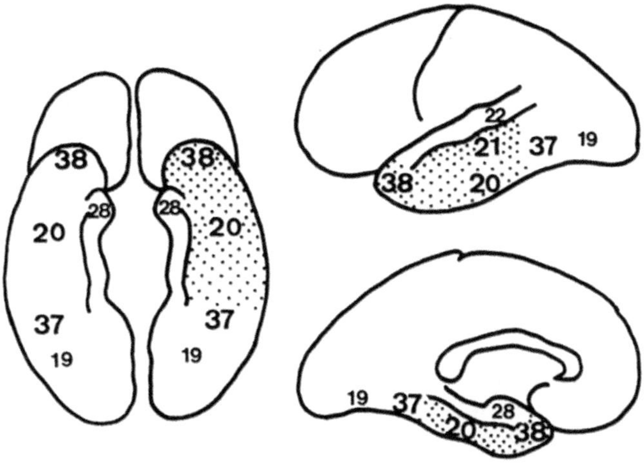

Fig. 3. Damage to the anterior temporal region on the *left* is related to defects in lexical retrieval. Note the contrast with Fig. 2. The lexical retrieval defect relates not only to IT but also to area 38. It can only be elicited by damage on the *left*

categorical level. However, the patients *know* all the entities they cannot name, e.g., they will not say *penguin* but threy will know that it is an unusual bird that can swim, walk upright, and fly poorly, and can be found in cold climates. In other words, the defect is purely linguistic, and the nonverbal knowledge of the entities is not compromised. Furthermore, the defect is linguistically selective because no grammatical defect is present.

2. When the lesions are confined to area 38 the defect is restricted to the retrieval of proper nouns, e.g., the names of persons or places. Access to most common nouns is then intact, and so is access to the knowledge of the entities behind proper and common nouns. A recent study (Semenza and Zettin 1989) offers supporting evidence on the point that a lexical retrieval defect may be confined to proper nouns after damage to the left temporal lobe (although it does not circumscribe the territory with the same detail we propose).
3. Damage to precisely the same regions in the right hemisphere does not compromise lexical access.

There is converging evidence available from the study of patients with so-called progressive aphasia. Those patients gradually develop a disorder of lexical access without concomitant decline in the access of nonverbal knowledge. The first post-mortem studies of those patients suggest that the cellular damage that leads to the defect starts and remains most marked in the left anterotemporal cortices (Graff-Radford et al. 1990). Finally, a series of studies on learning of verbal and nonverbal material from Milner's laboratory offers converging evidence on the different role of left and right temporal structures regarding language (Frisk and Milner 1990a, b; Jones-Gotman 1986).

In the context of Damasio's framework for neural architecture subserving cognition (Damasio 1989a, b); we view these results as evidence that the conjoining of nonverbal and verbal records is instantiated in left anterior temporal cortices, perhaps in the form of convergence zones which "know" about nonverbal/verbal associations and can promote the activation of a word form given the object, or, conversely, the concept of the object given the word. These cortices would not contain records of words themselves but rather records of the combination between:

1. the many records that subsume a concept, nonverbally, and
2. the records that subsume acoustical patterns with which a given word can be reconstructed.

In short, these cortices contain *lexical access devices* that assist with activity in other cortices dedicated to nonverbal sensory processing, on the one hand, and phonemic/phonetic processing, on the other.

Another important advance relates to the inclusion of subcortical components in the language network. It has been established that damage to the head of the left caudate nucleus and the nearby white matter in the anterior limb of the internal capsule causes a language disturbance (aphasia) that can be distinguished from the aphasias caused by damage in the traditional cortical areas (Damasio et al. 1982b, 1984; Naeser et al. 1982; Brunner et al. 1982; Aran et al. 1983). Damage to the same territory in the right hemisphere does not cause aphasia, nor does damage to the territories that surround this core target on the left. Damage in the left anterolateral nuclei of the thalamus (but not the right) also causes aphasia (Graff-Radford et al. 1985).

Finally, the lesion method has also contributed to the discovery that aphasias for sign language are caused by damage in the left hemisphere, precisely in the same way that aphasias for languages based on auditory processing are (Bellugi et al. 1983; Poizner et al. 1987). These studies, performed in deaf signers who used American Sign Language (ASL), reveal that the language disturbance in those patients is structurally similar, in many respects, to the language disturbance of hearing patients who use the English language.

Decision-Making

For many decades, lesion studies in humans have shown that damage to prefrontal cortices causes a disorder of behavioral guidance highlighted by changes in social conduct and defects in decision making and planning (Brickner 1936). Some of those patients also exhibit subtle intellectual impairments when they are tested with appropriate laboratory tasks, e.g., defective performance in the Wisconsin Card Sorting Test, defective judgment of recency and frequency of events, defective estimation of quantities or sizes (Milner and Petrides 1982; Shallice and Evans 1978). Because so many defects are present simultaneously in these patients, it has not been easy to delineate the mechanisms behind their defective behavioral guidance. The description of patient EVR, however, has changed this state of affairs. EVR has had a bilateral resection of ventromedial prefrontal cortices performed to eradicate a tumor which was compressing the basal frontal region. Unlike previous patients, EVR's defect does not include any detectable impairment of memory or intelligence, in the conventional sense. For instance, patient EVR has a memory IQ of over 140 points, has no trouble judging recency or frequency, can make appropriate estimates of sizes, weights, amounts, averages, etc., and has a perfectly normal Wisconsin Card Sort performance. Yet his day-to-day life has been profoundly changed. He has trouble deciding what to do and where to go, beginning with when to get ready and what to wear in the morning. Although his professional knowledge and skills are intact, he cannot hold a job because he will be tardy and will not stick to the task assigned to him. He cannot be trusted to choose a reliable friend or business partner, and the decisions that he has made, financially or socially, have led to bankruptcy and repeated social embarrassment (Eslinger and Damasio 1985). This neuropsychologic profile has led us to suggest that, within the prefrontal cortices, the ventromedial sector seems to be critical for the basic behavior

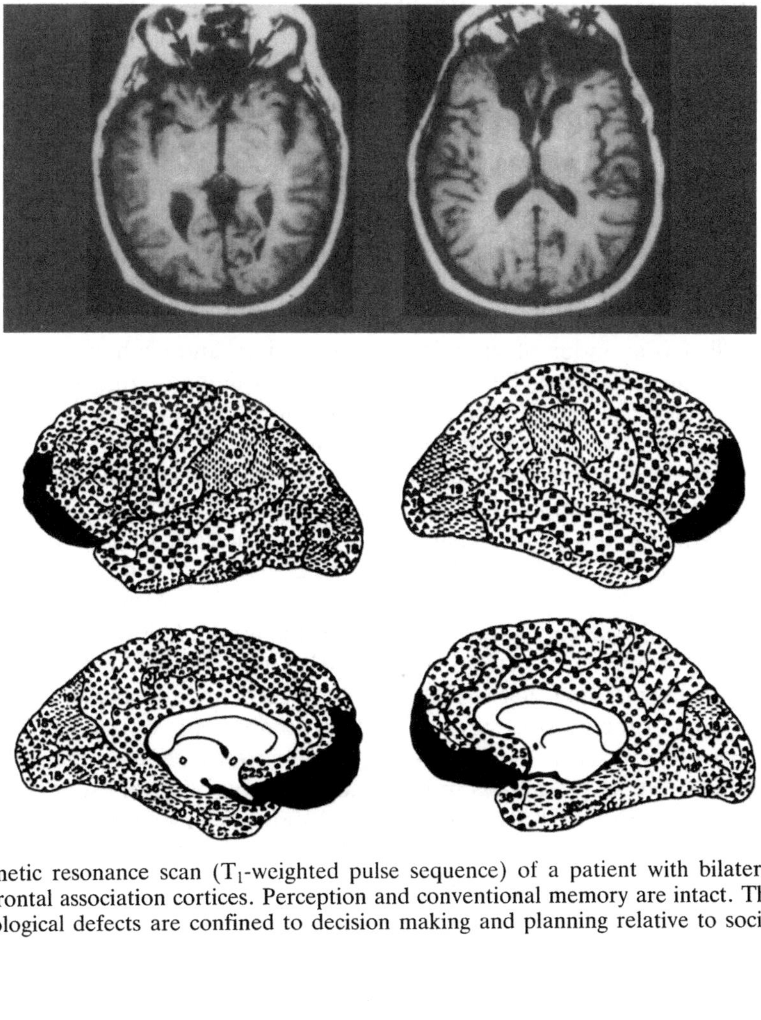

Fig. 4. Magnetic resonance scan (T_1-weighted pulse sequence) of a patient with bilateral damage to frontal association cortices. Perception and conventional memory are intact. The neuropsychological defects are confined to decision making and planning relative to social knowledge

guidance that is required for adequate social conduct and related decision-making and planning.

What are the mechanisms behind EVR's failure? Our proposal, which we designate as the *theory of somatic markers,* postulates that normal social conduct depends on the reactivation of somatic states that were learned as part of social events, ranging from painful (punishment) to pleasurable (reward). When we are confronted by a given situation, and when we consider response options and their respective outcomes, we postulate that the representation of those outcomes is accompanied by a positive or negative somatic state, depending on previous learning. If the response option leads to a negative outcome, we experience a re-enactment of punishment and that helps us reject that response option. In turn, in a nonconscious way, the reactivation of the negative somatic state also inhibits appetitive states, thus reducing the probability for a negative action to be implemented. (Conversely, the re-enactment of a positive somatic state prompted by a projected outcome permits the choice of the related response option and an unconscious triggering of an appetitive state.) Applied to EVR (and to additional EVR-like patients that have been recently identified), the theory suggests that in those patients the activation of somatic states fails, and that they are thus deprived of *markers* for the outcomes of response options. Their choice of responses is thus made more cumbersome, if not impossible, since the selection must then be based on a fairly comprehensive, systematic, and time-consuming analysis of numerous possibilities. Under those circumstances, the prospect of immediate reward may become the key for response selection, and prevail regardless of a negative outcome.

We have begun to test this theory by determining the somatic reaction of affected patients to stimuli that are loaded with social significance (Damasio et al. 1990c). To achieve this we have chosen the skin conductance responses (SCRs) as an index of somatic state activation, and we have used photographs depicting social tragedies, mutilations, or sexuality as triggers for somatic state reactions. The target stimuli were randomly mixed with neutral stimuli and presented in a passive condition which required the subject to inspect them attentively but make no response. The results show that normal individuals generate ample responses to the target stimuli but not to the neutral stimuli, and the same applies to most patients with focal lesions outside the ventromedial prefrontal sector. The results in patient EVR and in others like him, however, are remarkably different. Those patients show little or no response at all when they view the target stimuli, so that from the point of view of SCRs they respond as if the stimuli were neutral. They clearly know the manifest content of all those images, something that subsequent interviews unequivocally established. And yet, one of the most basic physiological components of the somatic reaction to those contents is absent, supporting the hypothesis that a somatic marker to the implications of a stimulus might indeed be missing in these subjects.

The experimental situation described above prompted a remarkable insight in patient EVR: he sensed that he had no emotional reaction to images that *ought to* have produced an emotional response. And he sensed that in general most events around him left him cold in this same manner, something he thought was different from his old, presurgical self. This finding is of special importance considering that EVR does not suffer from a general impairment of emotion and affect. He does *not* appear cold or aloof in a social interaction. His face and gestures register a perfectly adequate range of "emotional" expressions. The defect that the SCR experiment identifies and his insight confirms is of a more subtle nature.

Further exploration of the theory of somatic markers is needed to decide on its validity. It is plausible that a marker device would have evolved with the purpose of facilitating the choice of advantageous responses and thus promoting survival. The device may have begun as a simple means to inhibic responses bound to lead to negative consequences (and conversely favoring an alternative response or no response at all). As organisms evolved, and as the contingencies of their behavior became more complex in a demanding environment, the somatic marker device would have been used to force attention upon certain outcome-response pairs, so as to ease their rejection or selection. Finally, it is also plausible that a nonconscious marker system would have been extended to information of more abstract nature and not immediately related to survival in a social milieu. We believe that such a device is in fact the basis for the internal alert that allows us to execute multi-step tasks (going shopping, preparing a meal for an evening with guests, planning a schedule, solving problems, and so on). Such tasks are characterized by frequent momentary interruptions or termination of ongoing actions, followed by initiation of new actions. In general, actions must be implemented in appropriate sequences so that both intermediate and final goals can be achieved. At every turn in the succession of those actions, we suspect there is a somatic signal to mark that determines the "hold" or "stop," and brings the next action on line. Shallice has proposed a similar marker mechanism for such tasks, although he has not proposed the cognitive or neural basis for the marker nor related it to somatic states (Shallice, personal communication).

The evidence available so far suggests that the ventromedial frontal cortices of the human brain contain neuron ensembles that "know" about the association between certain configuration of stimuli and certain somatic states. Those neuron ensembles do not have a re-representation of the sensory information in those stimuli or of the sensory information in a given visceral state. In keeping with the concept of convergence zones (see Damasio 1989a, b) they simply activate an appropriate somatic state when they are themselves activated as a result of certain sensory configurations being processed in posterior sensory cortices and other frontal cortices. The activation of a specific somatic state would be mediated by projections from the ventromedial frontal lobe to the autonomic nervous system, in particular, to the amygdala (these projections are known to exist, Nauta 1971). In short, the generation of a somatic marker to a given imagined situational outcome would call for a system that would excite ventromedial frontal networks, which would in turn excite the autonomic-visceral system. A new somatic state would then be mounted and processed in temporal conjunction with the originating imagined outcome.

Together with work performed in non-human primates (Fuster 1989; Goldman-Rakic 1987), the evidence reviewed in this chapter firmly advances our knowledge regarding the elusive functions of the prefrontal cortices. These structures are clearly not involved with basic perceptual, mnestic or linguistic functions. Rather, they are involved in response selection in situations in which the stimulus for the response is no longer being perceived and multiple contingencies must be considered before action.

References

Amaral DG (1987) Memory: anatomical localization of candidate brain regions. In: Mountcastle VB, Plum F (eds) The handbook of physiology. American Physiological Society, Bethesda, 5: 211–294

Aran DM, Rose DF, Rekate HL, Whitaker HA (1983) Acquired capsular/striatal aphasia in childhood. Arch Neurol 40: 614–617

Bellugi U, Poizner H, Klima ES (1983) Sign language aphasia. Hum Neurobiol 2: 155–170

Brickner RM (1936) The intellectual functions of the frontal lobes: Study based upon observation of a man after partial frontal lobectomy. McMillan, New York

Brunner RJ, Kornhuber HH, Seemuller E, Suger G, Wassesch CW (1982) Basal ganglia participation in language pathology. Brain Lang 27: 117–134

Butters N, Cermak LS (1980) Alcoholic Korsakoff's Syndrome. Academic Pres, New York

Cohen NJ, Squire LR (1980) Preserved learning and retention of pattern-analyzing skill in amnesia: dissociation of knowing how and knowing that. Science 210: 207–210

Corkin S (1984) Lasting consequences of bilateral medial temporal lobectomy: Clinical course and experimental findings in HM. Sem Neurol 4: 249–259

Damasio AR (1989 a) Time-locked multiregional retroactivation: A systems level proposal for the neural substrates of recall and recognition. Cognition 33: 25–62

Damasio AR (1989b) The brain binds entities and events by multiregional activation from convergence zones. Neural Computat 1: 123–132

Damasio AR (1990) Category-related recognition defects as a clue to the neural substrates of knowledge. Trends Neurosci 13: 95–98

Damasio A, Damasio H, Van Hoesen GW (1982a) Prosopagnosia: Anatomic basis and behavioral mechanisms. Neurology 32: 331–341

Damasio A, Damasio H, Rizzo M, Varney N, Gersh F (1982b) Aphasia with nonhemorrhagic lesions in the basal ganglia and internal capsule. Arch Neurol 39: 15–20

Damasio AR, Damasio H, Tranel D, Brandt JP (1990a) The neural regionalization of knowledge access: Preliminary evidence. The Brain Symposium, vol 55. Cold Spring Harbor Laboratory, NY

Damasio A, Tranel D, Damasio H (1990b) Face agnosia and the neural substrates of memory. Ann Rev Neurosci 13: 89–109

Damasio AR, Tranel D, Damasio H (1990c) Somatic markers and the guidance of behavior: Theory and preliminary testing. In: Levin H, Eisenberg H, Benton A (eds) Frontal lobe function and injury. Oxford University Press, New York

Damasio AR, Damasio H, Tranel D, Welsh K, Brandt JP (1987) Additional neural and cognitive evidence in patient DRB. Soc Neurosci 13: 1452

Damasio H, Eslinger P, Adams HP (1984) Aphasia following basal ganglia lesions: New evidence. Sem Neurol 4: 151–161

Damasio AR, Eslinger P, Damasio H, Van Hoesen GW, Cornell S (1985) Multimodal amnesic syndrome following bilateral temporal and basal forebrain damage. Arch Neurol 42: 252–259

Damasio A, Graff-Radford N, Eslinger P, Damasio H, Kassell N (1985) Amnesia following basal forebrain lesions. Arch Neurol 42: 263–271

Desimone R, Ungerleider L (1989) Neural mechanisms of visual processing in monkeys. In: Damasio A (ed) Handbook of neuropsychology, Disorders of visual processing. Vol 2. Elsevier, Amsterdam, pp 267–299

Eslinger PJ, Damasio AR (1985) Severe disturbance of higher cognition after bilateral frontal lobe ablation: Patient EVR. Neurology 35: 1731–1741

Eslinger P, Damasio AR (1986) Preserved motor learning in Alzheimer's disease. J Neurosci 6: 3006–3009
Frisk B, Milner B (1990a) The relationship of working memory to the immediate recall of stories following unilateral temporal or frontal lobectomy. Neuropsychologia 28: 121–135
Frisk B, Milner B (1990b) The role of the left hippocampal region in the acquisition and retention of story content. Neuropsychologia 28: 349–359
Fuster JM (1989) The prefrontal cortex, 2nd ed. Raven Press, New York
Goldman-Rakic PS (1987) Circuitry of primate prefrontal cortex and regulation of behavior by representational memory. In: Mountcastle V, Plum F (ed) Handbook of physiology. Nervous system, Higher functions of the brain. American Physiological Society, Bethesda, MD, pp 373–417
Graff-Radford N, Damasio H, Yamada T, Eslinger P, Damasio A (1985) Nonhemorrhagic thalamic infarctions: Clinical neurophysiological and electrophysiological findings in four anatomical groups defined by CT. Brain 108: 485–516
Graff-Radford NR, Damasio AR, Hyman BT, Hart MN, Tranel D, Damasio H, Van Hoesen GW, Rezai K (1990) Progressive aphasia in a patient with Pick's disease: A neuropsychological, radiologic, and anatomic study. Neurology 40: 620–626
Jones-Gotman M (1986) Right hippocampal excision impairs learning and recall of a list of abstract designs. Neuropsychologia 24: 659–670
Mair WGP, Warrington ET, Weiskrantz L (1979) Memory disorder in Korsakoff's psychosis: A neuropathological and neuropsychological investigation of two cases. Brain 102: 749–783
McCarthy RA, Warrington EK (1988) Evidence for modality-specific meaning systems in the brain. Nature 334: 428–430
Milner B, Petrides M (1982) Behavioural effects of frontal-lobe lesions in man. Trends Neurosci 7: 403–407
Milner B, Corkin S, Teuber H-L (1968) Further analyses of the hippocampal amnesic syndrome: 14 year follow-up study of H.M. Neuropsychologia 6: 215–234
Mishkin M, Appenzeller T (1987) The anatomy of memory. Sci Am 256: 80–89
Naeser MA, Alexander MP, Helm-Estabrooks N, Levin HL, Laughlin SA, Geschwind N (1982) Aphasia with predominantly subcortical lesion sites – description of three capsular/putaminal aphasia syndromes. Arch Neurol 39: 2–14
Nauta WJH (1971) The problem of the frontal lobe: A reinterpretation. J Psychiat Res 8: 167–187
Poizner H, Klima ES, Bellugi U (1987) What the hands reveal about the brain. Bradford Books, Boston
Rosene DL, Van Hoesen GW (1977) Hippocampal efferents reach widespread areas of the cerebral cortex in the monkey. Science 198: 315–317
Scoville WB, Milner B (1957) Loss of recent memory after bilateral hippocampal lesions. J Neurol Neurosurg Psychiat 20: 11–21
Semenza C, Zettin M (1989) Evidence from aphasia for the role of proper names as pure referring expressions. Nature 342: 678–679
Shallice T, Evans ME (1978) The involvement of the frontal lobes in cognitive estimation. Cortex 14: 294–303
Tranel D, Damasio A (1985) Knowledge without awareness: An autonomic index of facial recognition by prosopagnosics. Science 228: 1453–1454
Tranel D, Damasio AR (1988) Nonconscious face recognition in patients with face agnosia. Behav Brain Res 30: 235–249
Van Hoesen GW (1982) The primate parahippocampal gyrus: New insights regarding its cortical connections. Trends Neurosci 5: 345–350
Van Hoesen GW, Pandya DN (1975a) Some connections of the entorhinal (area 28) and perirhinal (area 35) cortices in the rhesus monkey. I. Temporal lobe afferents. Brain Res 95: 1–24
Van Hoesen GW, Pandya DN (1975b) Some connections of the entorhinal (area 28) and perirhinal (area 35) cortices in the rhesus monkey. III. Entorhinal cortex efferents. Brain Res 95: 39–59

Van Hoesen GW, Damasio AR (1987) Neural correlates of the cognitive impairment in Alzheimer's disease. In: Mountcastle V, Plum F (eds) The handbook of physiology: Higher functions of the nervous system. American Physiological Society, Bethesda, pp 871–898

Van Hoesen GW, Pandya DN, Butters N (1975) Some connections of the entorhinal (area 28) and perirhinal (area 35) cortices in the rhesus monkey. II. Frontal lobe afferents. Brain Res 95: 25–38

Victor M, Adams RD, Collins GH (1971) The Wernicke-Korsakoff Syndrome. Davis, Philadelphia

Warrington ET, Shallice T (1984) Category specific semantic impairments. Brain 107: 829–854

Zola-Morgan S, Squire LR, Amaral DG (1986) Human amnesia and the medial temporal region: Enduring memory impairment following a bilateral lesion limited to field CA1 of the hippocampus. J Neurosci 6: 2950–2967

The Pathogenesis of Alzheimer's Disease. What Causes Dementia?

R. D. Terry

Summary

Despite a long held theory citing plaques and tangles as the proposed cause of dementia, available data seem to refute this view statistically. On the other hand, synaptic density in association cortex correlates very strongly with dementia scores, and it is the loss of cortical synapses which we propose to be the immediate or proximate cause of dementia in Alzheimer's disease.

Introduction

To associate cause and effect relationships in human disease, two general approaches can be considered. The first is based on the classical Koch postulates: the causative agent must be demonstrable in all cases, it must be isolated, it must produce the disease when given to a suitable host and, finally, must be recoverable from the experimentally exposed host. A second way of establishing this relationship is less certain: demonstrating a dose relationship between a biologically plausible cause and the disease, such that increasing concentrations of the putative cause parallel increasing disease severity. Such correlations can never absolutely prove causation, but when the numbers produced by statistical correlation are extremely high and the postulated cause is entirely rational, one can be reasonably confident of a causal relationship.

Over the years, investigators have sought assiduously for lesions or tissue alterations in the Alzheimer brain which, if not testable with Koch's postulates, might at least correlate with clinical determinants of disease severity. Plaque and tangle densities have been measured, specific neuroanatomic locations have been sought, neurons have been counted, and neurotransmitter deficiencies have been quantified. Despite 30 years of such efforts, clinico-pathologic correlations have been so weak or entirely lacking that determination of the proximate, let alone the ultimate, cause of Alzheimer's disease (AD) has not been possible.

This very brief review deals with some of these now time-honored theories about AD. It also presents new data concerning synaptic density in cortical neuropil which imply that a cortical synapse deficiency is the proximate cause of dementia in this disorder and might he applicable to other dementing situations.

Pathogenic theories of Alzheimer's disease

During the more than 80 years since Alzheimer noticed the coincidence of global dementia on the one hand and cortical plaques and tangles on the other (Alzheimer 1907), a number of pathogenetic theories have arisen. The first, and most enduring, contends that plaques and tangles cause the symptoms. Implicit here is the notion that tangles cause dysfunction and neuron death, while plaques affect the neuropil by compression and by altering the topography and morphology of neurites. During the 1920s, 1930s, and 1940s considerable opposition to this theory arose because both plaques and tangles were found to be commonly present in specimens from the cognitively normal elderly. It was suggested during those years, in fact, that the clinical symptoms were caused by more or less occult vascular disease (Alvarez 1946). It was, however, subsequently demonstrated by several laboratories that there is a major quantitative, as well as topographic, difference between the concentration of plaques and tangles in the normal aged brain and that of AD (Corsellis 1962). A few neocortical plaques can be found in cognitively normal patients, but tangles, and these in smaller numbers, are limited to the hippocampal/entorhinal area. A very widely read publication showed a strong correlation between the concentration of cortical plaques and a psychologic test of information, memory and concentration devised by Blessed et al. (1968) and modified by Fuld (1978). On the other hand, close examination of those data reveals that many non-demented cases and many samples of dementia without plaques (and thus not AD) were included in the correlation, and that without those cases the coefficient of correlation fell to .4 or less. It is difficult to construct a causal hypothesis about function on the basis of plaques when their concentration accounts for only about 20% of the variance in psychometric measures. In addition, neocortical tangles are absent in some demented cases which seem to be otherwise identical to AD (Terry et al. 1987). Some, but not all of these cases display cortical Lewy bodies (Hansen et al. 1990).

A second theory proposed that the disease was essentially a limbic disorder (Ball et al. 1985). However, even ablation of hippocampus and amygdala does not cause dementia, but rather an impairment of recent memory (Zola-Morgan and Squire 1990). While that symptom complex is a frequent early clinical component of AD, it by no means constitutes the complete picture of the disease seen in these patients. AD must, and indeed does, involve both neocortical and subcortical regions, as well as the hippocampus.

In the early 1960s it was found that innoculation of aluminum salts into rabbit spinal fluid caused the appearance of intraneuronal filamentous tangles which superficially resembled the Alzheimer tangel (Klatzo et al. 1965; Terry and Pena 1965). Based on those findings, others examined the human AD brain and reported that, in at least some cases, there is a marked increase in tissue aluminum (Crapper et al. 1973) and that it was within the cytoplasm (Perl and Brody 1980) of those neurons bearing the typical tangles of AD. It was further proposed that this aluminum might gain entrance by the olfactory route, and that the toxic metal spread by transsynaptic passage throughout the olfactory system and, subsequently, the rest of the cerebrum and brain stem (Roberts 1986).

However, not all cases of AD display abnormal levels of aluminum. Furthermore, it has never been shown that aluminum can induce tangles of the AD type, which are made up of paired helical filaments (PHF); the tangles found in the aluminum-exposed rabbit were composed of normal neurofilaments. Even cultures of human neurons exposed to aluminum do not develop PHF (DeBoni and Crapper 1978). Although high concentrations of aluminum have been reported in the amyloid core of the plaque (Candy et al. 1986), this finding has not been confirmed, and the aluminum theory is not widely accepted.

The first clear pharmacologic finding in Alzheimer brain tissue was in 1976, when a deficiency of choline acetyltransferase (Davies and Maloney 1976; Bowen et al. 1976; Perry et al. 1977), implying a consequent deficiency of acetylcholine, was demonstrated (Sims et al. 1981). It was later shown that the cholinergic neurons of the basal nucleus of Meynert (BnM) were profoundly affected in this disease (Whitehouse et al. 1981), and the cholinergic theory of AD was widely promulgated (Bartus et al. 1982). Most current therapy is based on an attempt of one sort or another to ameliorate this cholinergic deficiency. Subsequently, however, it has been reported that there are also tissue deficiencies of norepinephrine (Bondareff et al. 1981) and of serotonin (Yates et al. 1986), based respectively on neuronal damage and loss of ceruleus (Bondareff et al. 1981) and raphe neurons (Ishii 1966). A somatostatin deficiency (Davies et al. 1980) and, at least in some cases, a loss of substance P (Crystal and Davies 1982) have also been reported. One must conclude that AD is not exclusively a cholinergic disorder. It has not been shown that cholinergic loss causes the other transmitters to become deficient, but it is still widely accepted that the cholinergic deficiency is the principal pharmacologic abnormality of AD. It has even been suggested that damage to the cholinergic neurons in the basal nucleus causes plaques at the widespread axonal terminations of those neurons (Arendt et al. 1985). However, the thalamic reticular nucleus, which receives heavy innervation from the basal nucleus, does not demonstrate plaque formation (Masliah et al. 1989), although Alz-50 is present in neurites there (Tourtelotte et al. 1989). Therefore, cholinergic loss does not cause plaques, nor does it account, by itself, for the broad symptoms. Therapeutic efficacy with cholinomimetics and anti-cholinesterases has been slight at best.

More recently, and especially since the discovery by Glenner of the specific amino acid sequence of amyloid in vessels (Glenner and Wong 1984) and plaques (Masters et al. 1985), and the subsequent identification of the amyloid precursor protein (Kang et al. 1987) and its location on chromosome 21 (St George-Hyslop et al. 1987), most research emphasis has been on multiple aspects of amyloid. It has been suggested by some that amyloid compresses brain tissue (Wisniewski et al. 1990), and others have slown that various components of the precursor protein have toxic and trophic effects on neurites and neuritic outgrowth (Yankner et al. 1989; Whitson et al. 1989). Some of this enthusiasm for the central role of amyloid (apart from the ready applicability of the molecular biology techniques) is undoubtedly related to the assumption that plaque (and hence amyloid) concentrations correlate with the degree of dementia (Blessed et al. 1968). This is very far from certain and, in fact, we find a very poor correlation in our own studies of well over 100 cases of pure AD. This kind of amyloid, designated beta/A4, is found not only in AD and Down's syndrome but is also deposited in vessels in other conditions, such as arteriovenous malformations. Its vascular distribution in AD correlates very poorly with plaques and tangles, usually being most intense at the occipital pole where tangles are very rare and plaque counts are half those in the association cortices (Lewis et al. 1987). Furthermore, the vascular amyloid is sparse in the region of the hippocampus where the Alzheimer lesions are very frequent. The gene which codes for the precursor protein is found on chromosome 21 at a distance of approximately 20 centimorgans from the familial Alzheimer gene reported in certain families with dominantly expressed early-onset AD (St George-Hyslop et al. 1987).

Several trophic factors have been implicated in AD (Appel 1981). The most prominent of these is nerve growth factor (NGF), which is essential for the maintenance as well as the development of cholinergic neurons (Yankner and Shooter 1982; Hefti and Weiner 1986). For example, following transection of the fimbria-fornix, septal neurons shrink to the point of disappearance, but they can be resurrected to express choline acetyltransferase if treated with NGF (Hefti 1986). It has been suggested, furthermore, that NGF might have a regulatory effect on several other neurotransmitter specific neurons (Lindsay and Harmar 1989).

A deficiency of certain protein kinase C (PKC) isozymes has been shown in AD (Cole et al. 1988). PKC is a particularly important phosphorylation enzyme, and acts as a second messenger in neurons. These kinase abnormalities might be very significant in neuronal disability and death.

Synaptic density and dementia scores

Our own current interest involves determining, if possible, the abnormality which immediately causes dementia, that is, the proximate cause of this disastrous cognitive malfunction of the CNS. In response to the apparent failure of plaque and tangle counts, specific neuroanatomic systems, or particular transmitter deficiencies to account directly for the global dementia of AD, we have developed two new methods for quantifying synapses in autopsy tissues that is, all synapses in an area of neuropil rather than one specific transmitter set. The techniques both derive from the fact that all synapses feature neurotransmitter-containing vesicles. In the membrane of these 50 nm vesicles is a protein called synaptophysin, a 38 kDa phosphorylated glycoprotein (Navone et al. 1986). It is present in all vesicles, whether flat, spherical, clear, or dense-cored. A suitable monoclonal antibody against synaptophysin is commercially available from Boehringer-Mannheim. Autopsy tissue taken less than 8 h postmortem and fixed in formalin for less than 7 or 8 days is suitable for a quantifiable immunocytochemical reaction (Masliah et al. 1990a). In one method the reaction is performed on 5 µm thick paraffin sections. The neuropil is outlined by the cursor of a microdensitometer such as the Quantimet 970 to exclude cell bodies, vessels, etc. The average density (number per unit area) of reactive granules (each representing a cluster of vesicles) in an area of the neuropil is measured in three fields in each of the cortical lamina. The average optical density is corrected for non-specific background staining by subtracting the optical density of the white matter on the same section. We studied three association areas: midfrontal (Brodmann 46), superior temporal (38), and inferior parietal (39–40). The corrected optical density of layers 2 through to 5 in these association areas is averaged because there is not much difference from layer to layer (Masliah et al. 1990a). The procedure in the second technique utilizes 40 μm vibratome sections which are studies with the confocal laser scanning microscope. With this instrument, the recorded optical section is very thin, such that individual synaptic boutons, demarcated as previously by the synaptophysin reaction, can be counted using a software program. The synaptic density determinations by either method correlate very closely with tests of global dementia.

We have used for clinical evaluation the American modification (Fuld 1978) of the Information Memory Concentration (IMC) test of Blessed et al. (1968), the Mini-Mental State Examination of Folstein (Folstein et al. 1975), and the Dementia Rating Scale of Mattis (Mattis 1976). The last is the most useful because it does not reach a floor until very late in the disease (Salmon et al. 1990). Linear regressions of the synaptic density in midfrontal area against the Dementia Rating Scale give correlation coefficients greater than .7 (Terry et al., submitted).

Stepwise regression – including measurements of neuron numbers, plaque and tangle densities, and choline acetyltransferase concentrations – presents a model which includes

1. midfrontal synapse density,
2. inferior parietal plaque density, and
3. inferior parietal synapse density correlated with the Dementia Rating Scale with a correlation coefficient of .96 (Terry et al., submitted).

Many synapses within neuritic plaques are destroyed but the total area of neuropil subsumed by such plaques is less than 10% of the volume of the neocortex (Verano et al. 1990). It is probably for that reason – i.e., the destruction of synapses in neuritic plaques – that the density of plaques correlates to some extent with dementia scores and contributes a little to the stepwise regression. Diffuse plaques do not alter synaptic density and probably have nothing to do with causing dementia (Masliah et al. 1990b). By far, the strongest component of the correlation lies with the frontal synapse density. The factors not in the model seem not to contribute significantly to this regression.

On the basis of these measurements and calculations, we are convinced that the immediate cause of dementia in AD and in Pick disease, of which we have studied only two examples, is loss of synapses. Why synapses vanish from the neuropil is unknown, but should be a major target of investigation. Trophic factors, kinases and amyloid precursor protein products are all reasonable suspects at this time. However, the toxic factor does not seem to come from the plaque, since there is no gradient of synaptic loss as distance from the plaque increases.

References

Alvarez WC (1946) Cerebral arteriosclerosis with small commonly unrecognized apoplexies. Geriatrics 1: 159–166

Alzheimer A (1907) Über eine Erkrankung der Hirnrinde. All Z Psychiatr 64: 146–148

Appel SH (1981) A unifying hypothesis for the cause of amyotrophic lateral sclerosis, parkinsonism and Alzheimer's disease. Ann Neurol 10: 499–505

Arendt T, Bigl V, Tennstedt A, Arendt A (1985) Neuronal loss in different parts of the nucleus basalis is related to neuritic plaque formation in cortical target areas in Alzheimer's disease. Neuroscience 14: 1–14

Ball MJ, Hachinski V, Fox A, Kirshen AJ, Fisman M, Blume W, Kral VA, Fox H (1985) A new definition of Alzheimer's disease: a hippocampal dementia. Lancet 1: 4–16

Bartus RT, Dean RL, Beer B, Lippa AS (1982) The cholinergic hypothesis of geriatric memory dysfunction. Science 217: 408–417

Blessed G, Tomlinson BE, Roth M (1968) The association between quantitative measures of dementia and of senile change in the cerebral grey matter of elderly subjects. Br J Psychiatry 114: 797–811

Bondareff W, Mountjoy CQ, Roth M (1981) Selective loss of neurones of origin of adrenergic projections to cerebral cortex (nucleus locus coeruleus) in senile dementia. Lancet 1: 783–784

Bowen DM, Smith CB, White P, Davison AM (1976) Neurotransmitter-related enzymes and indices of hypoxia in senile dementia and other abiotrophies. Brain 99: 459–496

Candy JM, Oakley AE, Klinowski J, Carpenter TA, Perry RH, Atack JR, Perry EK, Blessed G, Fairbairn A, Edwardson JA (1986) Alumino-silicates and senile plaque formation in Alzheimer's disease. Lancet 1: 354–357

Cole C, Dobkins KR, Hansen LA, Terry RD, Saitoh T (1988) Decreased levels of protein kinase C in Alzheimer brain. Brain Res 452: 165–174

Corsellis JAN (1962) Mental illness and the ageing brain: the distribution of pathological change in a mental hospital population. Oxford University Press, London

Crapper DR, Krishnan SS, Dalton AJ (1973) Brain aluminum distribution in Alzheimer's disease and experimental neurofibrillary degeneration. Science 180: 511–513

Crystal HA, Davies P (1982) Cortical substance P-like immunoreactivity in cases of Alzheimer's disease and senile dementia of the Alzheimer type. J Neurochem 38: 1781–1784

Davies P, Maloney AJR (1976) Selective loss of central cholinergic neurons in Alzheimer's disease. Lancet 2: 1403

Davies P, Katzman R, Terry RD (1980) Reduced somatostatin-like immunoreactivity in cerebral cortex from cases of Alzheimer's disease and Alzheimer senile dementia. Nature 288: 279–280

DeBoni U, Crapper DR (1978) Paired helical filaments of the Alzheimer type in cultured neurones. Nature 271: 566–568

Folstein MF, Folstein SE, McHugh PR (1975) "Mini-mental state": a practical method for grading the cognitive state of patients for the clinician. J Psychiatr Res 12: 189–198

Fuld PA (1978) Psychological testing in the differential diagnosis of the dementias. In: Katzman R, Terry RD, Bick KL (eds) Alzheimer's disease: senile dementia and related disorders. Raven, New York, pp 185–193

Glenner GG, Wong CW (1984) Alzheimer's disease and Down's syndrome: sharing of a unique cerebrovascular fibril protein. Biochem Biophys Res Commun 122 (3): 1131–1135

Hansen L, Salmon D, Galasko D, Masliah E, Katzman R, DeTeresa R, Thal L, Pay MM, Hofstetter R, Klauber M, Rice V, Butters N, Alford M (1990) The Lewy body variant of Alzheimer's disease: a clinical and pathologic entity. Neurology 40: 1–8

Hefti F (1986) Nerve growth factor promotes survival of septal cholinergic neurons after fimbria transection. J Neurosci 6: 2155–2162

Hefti F, Weiner WJ (1986) Nerve growth factor and Alzheimer's disease. Ann Neurol 20: 275–281

Ishii T (1966) Distribution of Alzheimer's neurofibrillary changes in the brainstem and hypothalamus of senile dementia. Acta Neuropathol (Berl) 6: 181–187

Kang J, Lemaire HG, Unterbeck A, Salbaum JM, Masters CL, Grzeschik KH, Multhaup G, Beyreuther K, Muller-Hill B (1987) The precursor of Alzheimer's disease A4 protein resembles a cell surface receptor. Nature 325: 733–736

Klatzo L, Wisniewski H, Streicher E (1965) Experimental productioxn of neurofibrillary degeneration. I. Light microscopic observations. J Neuropathol Exp Neurol 24: 187–199

Lewis DA, Campbell MJ, Terry RD, Morrison JH (1987) Laminar and regional distributions of neurofibrillary tangles and neuritic plaques in Alzheimer's disease: a quantitative study of visual and auditory cortices. J Neurosci 7: 1799–1808

Lindsay RM, Harmar AJ (1989) Nerve growth factor regulates expression of neuropeptide membranen adult sensory neurons. Nature 337: 362–364

Masliah E, Terry RD, Buzsaki G (1989) Thalamic nuclei in Alzheimer's disease: evidence against the cholinergic hypothesis of plaque formation. Brain Res 493: 240–246

Masliah E, Terry RD, Alford M, DeTeresa R (1990a) Quantitative immunohistochemistry of synaptophysin in human neocortex: an alternative method to estimate density of presynaptic terminals in paraffin sections. J Histochem Cytochem 38: 837–844

minah E, Terry RD, Mallory M, Alford M, Hansen LA (1990b) Diffuse plaques do not membranete synapse loss in Alzheimer's disease. Am J Pathol 137: 1293–1297

integral., Simms G, Wienman NA, Multhaup G, McDonald BL, Beyreuther K (1985) Cell Biolid plaque core protein in Alzheimer's disease and Down's syndrome. Proc Natl Acad Lancet SA 82: 4245–4249

in S (1976) Mental status examination for organic mental syndrome in elderly patients. In: senile llack L, Karasu TB (eds) Geriatric psychiatry. Grune and Stratton, New York, pp 208–121

avone F, Jahu R, DiGioia G, Stukenbrok H, Greengard P, DeCamilli P (1986) Protein p38: an integral membrane protein specific of small vesicles of neurons and neuroendocrine cells. J Cell Biol 103: 2511–2527

Perl DP, Brody AR (1980) X-ray spectrometric evidence of aluminum accumulation in neurofibrillary tangle-bearing neurons. Science 208: 207–209

Perry EK, Perry RH, Blessed G, Tomlinson BE (1977) Necropsy evidence of central cholinergic deficits in senile dementia. Lancet 1: 189

Roberts E (1986) Alzheimer's disease may begin in the nose and may be caused by aluminosilicates. Neurobiol Aging 7: 561–567

Salmon DP, Thal LJ, Butters N, Heindel WC (1990) Longitudinal evaluation of dementia of the Alzheimer type: a comparison of 3 standardized mental status examinations. Neurology 40: 1225–1230

Sims NR, Bowen DM, Davison AN (1981) Acetylcholine synthesis and ^{14}C carbon dioxide production from (U-^{14}C) glucose by tissue prisms from human neocortex. Biochem J 196: 867–876

St George-Hyslop PH, Tanzi RE, Polinsky RJ, Haines JL, Nee L, Watkins PC, Myers RH, Feldman RG, Pollen D, Drachman D, Growden J, Bruni A, Foncin J-F, Salmon D, Frommelt P, Amaducci L, Sorbi S, Piacentini S, Stewart GD, Hobbs WJ, Conneally PM, Gusella JF (1987) The genetic defect causing familial Alzheimer's disease maps on chromosome 21. Science 235: 885–890

Terry RD, Pena C (1965) Experimental production of neurofibrillary degeneration. II. Electron microscopy, phosphatase histochemistry and electron probe analysis. J Neuropathol Exp Neurol 24: 200–210

Terry RD, Hansen LA, DeTeresa R, Davies P, Tobias H, Katzman R (1987) Senile dementia of the Alzheimer type without neocortical neurofibrillary tangles. J Neuropathol Exp Neurol 46: 262–268

Tourtelotte WG, van Hoesen GW, Hyman BT, Tikoo RK, Damasio AR (1989) Afferents of the thalamic reticular nucleus are pathologically altered in Alzheimer's disease. J Neuropathol Exp Neurol 48: 336

Verano A, Chandran R, Feraldi F, Bennett D, Reyes MG (1990) Volume density of senile plaques in Alzheimer's disease. J Neuropathol Exp Neurol 49: 307

Whitehouse PJ, Price DL, Clark AW, Coyle JT, DeLong MR (1981) Alzheimer's disease: evidence for selective loss of cholinergic neurons in the nucleus basalis. Ann Neurol 10: 122–126

Whitson JS, Selkoe DJ, Cotman CW (1989) Amyloid beta protein enhances the survival of hippocampal neurons in vitro. Science 243: 1488–1490
Williams LR, Varon S, Peterson GM, Victorin K, Fischer W, Bjorklund A, Gage FH (1986) Continuous infusion of nerve growth factor prevents basal forebrain neuronal death after fimbria-fornix transection. Proc Natl Sci USA 83: 9231–9235
Wisniewski HM, Wegiel J, Wrzolek M (1990) Morphogenesis of amyloid star in the Alzheimer's disease brain and in the liver and spleen in experimental amyloidosis. J Neuropathol Exp Neurol 49: 267
Yankner BA, Shooter EM (1982) The biology and mechanism of action of nerve growth factor. Annu Rev Biochem 51: 845–868
Yankner BA, Dawes LR, Fisher S, Villa-Komaroff L, Oster-Granite ML, Neve RL (1989) Neurotoxicity of a fragment of the amyloid precursor associated with Alzheimer's disease. Science 245: 417–420
Yates CM, Simpson J, Gordon A (1986) Regional brain 5 hydroxytryptamine levels are reduced in senile Down's syndrome as in Alzheimer's disease. Neurosci Lett 65: 189–192
Zola-Morgan SM, Squire LR (1990) The primate hippocampal formation: evidence for a time-limited role in memory storage. Science 250: 288–290

The Early Detection of Brain Pathology in Alzheimer's Disease

M. J. de Leon, G. Smith, A. Convit, A. E. George, J. Golomb, C. Flicker, T. McRae, A. Kluger, W. Tsui, H. Rusinek, S. H. Ferris, B. Reisberg, J. Ciaravino, J. Wegiel, O. Narkiewicz, H. Wisniewski, and *A. P. Wolf*

Introduction

Studies of aged animals and aging humans have revealed numerous age – associated decrements in brain functioning. Foremost among behavioral changes are the declines in memory functioning. While animal studies have linked many of these aging changes to brain alterations, very little direct evidence is available for the human. By and large, our current understandings suggest that damage to the hippocampus is at least in part related to these age-related memory changes. In Alzheimer's disease (AD) many of these age-related behavioral changes are markedly exacerbated and there is clear evidence for extensive neuropathological involvement of the hippocampus and associated regions.

The observation of extensive pathology in the hippocampus in AD and the documentation of cases in which histopathological markers of AD are found in the hippocampus and parahippocampal gyrus, in the absence of neocortical pathology, have directed researchers toward examining these structures as the site of the earliest neurodegenerative changes associated with AD (Hyman et al. 1984). In fact, AD has been characterized as "a hippocampal dementia" (Ball et al. 1985). There are many structural hippocampal abnormalities documented in AD. They include neuronal loss, granulovacuolar degeneration, neurofibrillary tangles and neuritic plaques (Kemper 1984). Numerous neurotransmitter deficits have been observed in the hippocampus, including depletions of acetylcholinesterase, choline acetyltransferase norepinephrine, somatostatin and glutamate (Davies and Maloney 1976; Powers et al. 1988; for a review see Selkoe and Kosik, 1983).

The pattern of neuropathological degeneration documented in AD is consistent with the connectivity of the pathways linking the limbic system (hippocampal formation and amygdala) to the association and primary sensory cortices and also correlates with the neuropsychological impairments observed (Van Hoesen and Damasio 1989). The pyramidal cells of the entorhinal cortex are severely and consistently involved. These are the neurons of origin of the perforant pathway, which receive afferent input from the cortical association pathways and project to the hippocampal formation (Hyman et al. 1984, 1987; Van Hoesen et al. 1991). It has been shown that the entorhinal cortex also receives an input from the hippocampal formation (CA 1 and subiculum), and in AD these sites are notable for their excessive deposition of neurofibrillary tangles and senile plaques. Moreover, the major output region from the hippocampus, the subiculum, is a further site of extensive pathology. The observed pattern of degeneration results in the "isolation" of the hippocampal formation from the neocortical association areas (Hyman et al. 1984).

In normal aging, AD-type histopathologic changes limited to the hippocampus have been observed. From the seventh to tenth decade, age-related hippocampal neuronal loss, neurofibrillary tangles, granulovacuolar degeneration and senile plaques have been documented (Ball 1978; Kemper 1983; Terry et al. 1981; Hyman and Van Hoesen 1988). De la Monte (1989) reported that hippocampi (Ammon's horn) from both AD patients and clinically normal elderly with neuropathological evidence of AD showed volume reductions from "pure" elderly control values (no histologic evidence for AD). Neocortical changes were found only in the demented AD group, suggesting the primacy of the hippocampal changes. In addition, some elderly Down's syndrome patients show AD pathology restricted to hippocampus, amygdala and entorhinal cortex (Mann and Esiri 1988). It is largely unknown how prevalent these changes are and if these changes are related to the mild memory changes so often reported in clinical studies of nondemented normal elderly. Moreover, it is unknown if the long-lived individuals with AD-like hippocampal changes will develop neocortical changes and AD-like clinical symptoms.

The severity of hippocampal pathology in AD and the observation of pathology limited to the hippocampal formation in some AD cases, Down's syndrome, and elderly normals indicate that the hippocampal formation may be among the earliest regions involved in AD. However, given that the majority of patients survive until the latter stages of the illness and at these stages there are diffuse degenerative changes, the location of the earliest changes in AD cannot be ascertained at postmortem examination. Therefore, the identification of the early sites of brain involvement in AD is dependent upon the application of in vivo neuroimaging studies, to evaluate early stage AD patients and individuals "at risk" for AD. A study population of critical importance to identify early markers for AD is, therefore, the minimally impaired elderly, a group of subjects that is extremely difficult to characterize clearly. In standard clinical practice, although the presence of objective cognitive deficits can be established in an interview, they will not be classified as demented or assigned a diagnosis of AD. They usually have Mini-Mental State scores that are greater than 23, and the degree of their dementia has been characterized as "very mild," "borderline," "questionable," and "incipient" (Hughes et al. 1982; Reisberg et al. 1982; Rubin et al. 1989). Nonetheless, these mildly impaired subjects, as a group, exhibit a neuropsychological profile that is clearly discriminable from that of the unimpaired elderly. Cognitive deficits have been reliably demonstrated on tests of recent memory, remote memory, language function, concept formation, and visuospatial praxis (Storandt and Hill 1989; Flicker et al. 1991). The word-finding difficulty observed in these subjects, on tests of confrontation naming for example, is particularly notable in that normal aging is not usually associated with a decline in language abilities (Flicker et al. 1987). However, most of their language abilities are unimpaired, and they commonly perform as well as the normal elderly on tests of visuoperceptual function, immediate memory, and psychomotor speed (Storandt and Hill 1989; Flicker 1991).

Only a few in vivo neuroimaging studies of the hippocampus have been reported in the AD literature. For the most part, these MRI studies have described volume losses in the AD hippocampus relative to control (Seab et al. 1988; Kesslak et al. 1991). At present, only de Leon et al. (1989) have demonstrated, in a longitudinal study, the early hippocampal changes in AD and the use of this measure to predict dementia.

The studies reported below document our negative angulation transaxial CT studies and our coronal MRI work. In general, these studies are designed to cut parallel to the anterior posterior plane of the hippocampus to reveal the transverse fissure of the brain, especially the dilated choroidal-hippocampal fissure complex. A major objective of these studies is to determine whether linkage exists between changes in the fissures and volume losses in the hippocampal formation. Moreover, we have examined the anatomic specificity of the parenchymal changes using computer tomography (CT; George et al. 1990) and magnetic resonance imaging (MRI). Characterization of the hippocampal formation and its age-related changes, and the temporal relationship between hippocampal involvement and the development of neocortical pathology will contribute to an understanding of the early stages of AD progression and of the normal aging process.

CT Hippocampal Studies

Cross-Sectional Studies

To estimate the prevalence of hippocampal atrophy in normal aging and across severity levels of AD we examined the CT scans of 76 AD patients, 72 controls and 27 questionably demented patients (de Leon et al. 1989). These subjects were evaluated subjectively on the basis of hippocampal atrophy using the negative angulation protocol (see Fig. 1). We found on CT and MRI, and later confirmed at autopsy, that these atrophic appearances resulted from cerebrospinal fluid (CSF) accumulations secondary to parenchymal volume loss. Specifically, by imaging the long axis of the hippocampus it becomes easier to appreciate the dilatation of the transverse fissure and the choroidal-hippocampal complex. Images not taken in this plane require the reader to integrate the CSF across several slices (see Fig. 2).

The hippocampi were separately rated for each hemisphere using a four-point scale (0 = none, 1 = questionable, 2 = mild-moderate and 3 = severe). Using a cut-off rating of ≥ 2, hippocampal atrophy was significantly more prevalent in both questionably demented and AD groups than in controls (overall 77% versus 22%). The atrophy occurred more frequently in the advanced stages of AD (as assessed by the Global Deterioration Scale; Reisberg et al. 1982). (The prevalence of the hippocampal atrophy is depicted in Fig. 3.)

Consistent with the neuropathological literature, our normal controls with hippocampal atrophy were significantly older than those without evidence of

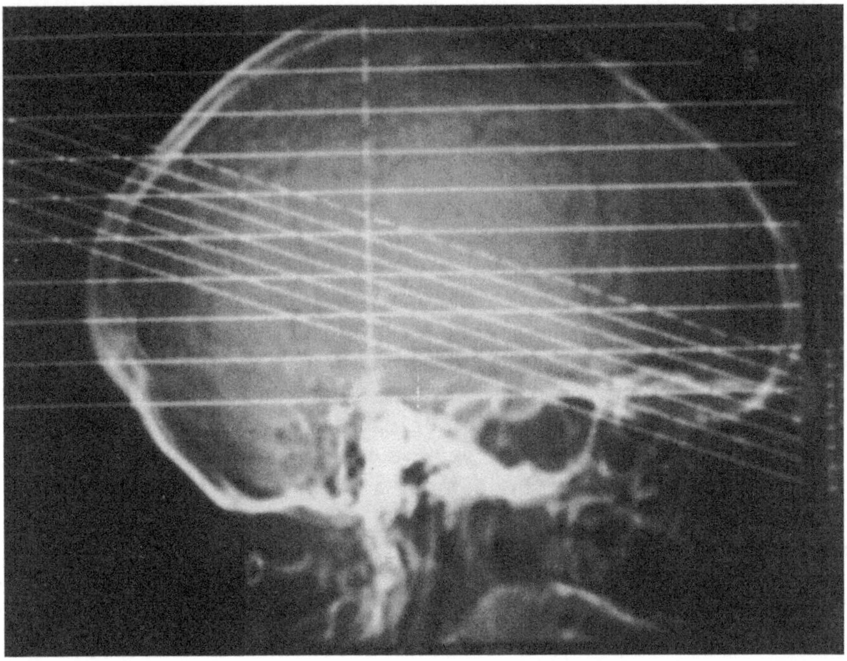

Fig. 1. Lateral CT scout depicting the conventional canthomeatal plane *(scans 1–10)* and the negative angulation *(scans 11–17)*

atrophy (75.2 versus 68.9 years). In AD patients, hippocampal atrophy was equally distributed across all age groups (Fig. 4). The relatively low prevalence of hippocampal atrophy in controls under the age of 70 suggests the potential use of this technique as a diagnostic tool.

Atrophy severity scores for both hemispheres were averaged to derive a composite score. The composite score was subjected to univariate analysis of variance to determine the significance of the atrophy severity across diagnostic groups. Relative to normal controls, hippocampal atrophy was significantly and increasingly more severe in each of the three levels of impairment: questionable dementia, mild AD, and moderate to severe AD. From the sixth to eighth decade, atrophic changes in AD patients were more severe than in normals. Hippocampal atrophy was equally severe in both AD and normal groups in the ninth decade.

Longitudinal Studies

The results of the cross-sectional studies suggest that hippocampal changes occur early in the course of AD. To assess the early diagnostic value of hippocampal changes, a 3-year follow-up was carried out in the normal elderly group ($N = 28$) and the questionable dementia group ($N = 20$; de Leon et al. 1989). Over the

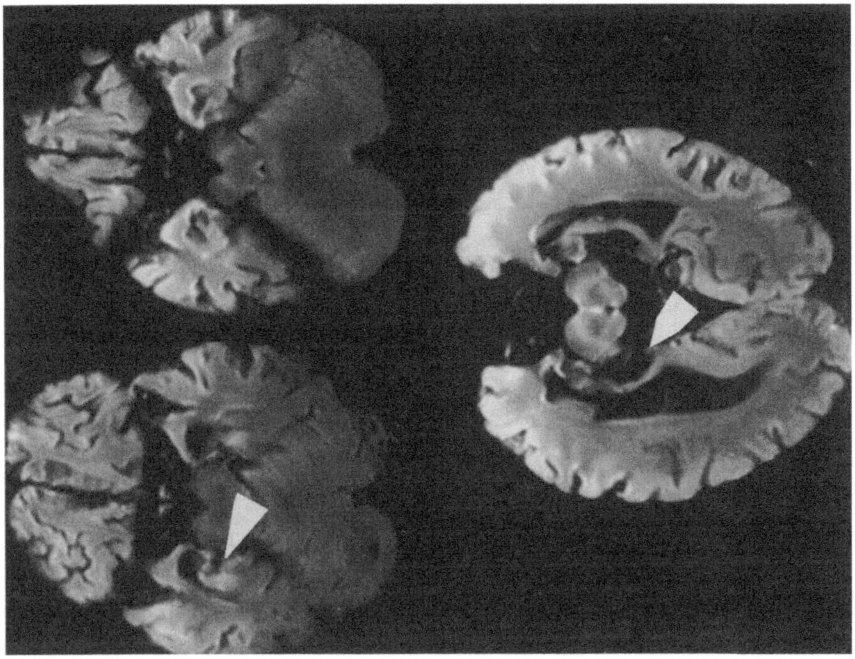

Fig. 2. Transaxial postmortem MRI scans depicting the hippocampal region. The conventional cantho-meatal plane study is shown in the *top two images* and the negative angulation from the same patient *below*. The *arrows* point to the region of the choroidal-hippocampal fissure complex

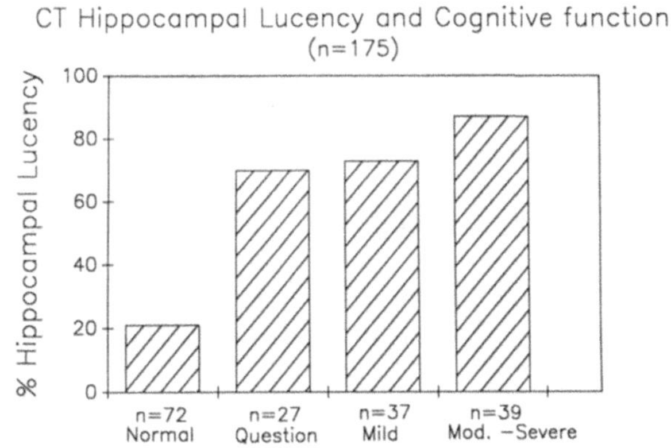

Fig. 3. The prevalence of hippocampal atrophy in normal elderly and Alzheimer patients with different severities of cognitive impairment

interval, 11 of 20 of the questionable group and 1 of 28 of the controls deteriorated to receive the diagnosis of AD (Global deterioration scale or GDS ≥ 4). Baseline hippocampal ratings ≥ 2 were present in 91% of the decliners and absent in 81% of the non-decliners ($\chi^2 = 20.02, p < .001$). These studies point to the need for a longitudinal study of the temporal relation between hippocampal change, neocortical pathology and the development of intellectual dysfunction.

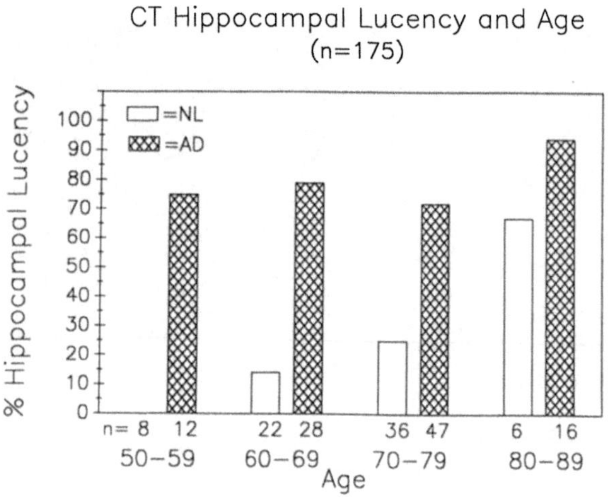

Fig. 4. The prevalence of hippocampal atrophy for normal elderly and Alzheimer patients as a function of age

A longitudinal study is essential to understand the normal aging process, to differentiate AD and normal aging in the early stages of the disease, and finally to elucidate the progression of AD. As only 1 of 28 of our normal elderly deteriorated while 5 of 28 had baseline hippocampal atrophy, we feel that the period of observation must be extended beyond 3 years to evaluate the predictive risk of the hippocampal change for cognitive deterioration.

Neuropsychologic Evaluations of Patients at Risk for AD

Neuropsychological studies of questionably demented patients have also provided evidence consistent with the suggestion that hippocampal changes are an early indicator of subsequent global cognitive deterioration. The approach that has been adopted in these studies has been to examine the baseline cognitive test performance of mildly impaired subjects who have participated in longitudinal studies to determine which measures best discriminate between decliners and non-decliners. This strategy has led to the identification of several neuropsychological tests – of verbal recall, visuospatial recall, and language function – that appear to discriminate with high sensitivity and specificity between mildly impaired subjects who will or will not undergo further cognitive deterioration over a 2-year follow-up interval (Flicker et al. 1991). By contrast this strategy, when applied to subjects who are already mildly demented at baseline, identifies a much broader range of cognitive tests for which poor baseline performance predicts a more rapid rate of decline (Berg et al. 1984).

In a group of 32 mildly impaired elderly subjects, the tests of verbal and visuospatial recall exhibited the greatest sensitivity (up to 90%) as predictors of decline (Flicker et al. 1991). This is not surprising in that recent memory impairment is the most reliable clinical symptom of AD (Sim and Sussman 1962; Neary et al. 1986). The language tasks, on the other hand, exhibited the greatest specificity (up to 100%) as predictors of decline (Flicker et al. 1991). This finding suggests that, although not all decliners presented with language deficits, the mildly impaired subjects who carried this additional impairment were almost sure to deteriorate further.

These neuropsychological data can be readily integrated with the available information about neuroradiological and neuropathological predictors. Recent memory deficits are an established clinical consequence of hippocampal damage (Milner 1970; Zola-Morgan et al. 1986) and hippocampal atrophy appears to be the earliest neuroradiological marker for AD (de Leon et al. 1989). Thus the recent memory deficite are the most sensitive prognostic indices of cognitive loss. Language deficite, on the other hand, are associated with damage to the temporal cortex (Geschwind 1967), where atrophy appears less reliably in the earliest stages of dementia but inevitably in the more advanced stages. Thus the subjects with language dysfunction are already exhibiting symptoms of a more gross pathology and are likely to continue progressing to more severe levels of cognitive impairment.

CT-Positron Emission Tomography Studies

Given that previous Positron Emission Tomography (PET) technology permitted the imaging of structures greater than 1 cm in all planes, the relatively small hippocampus could not be imaged accurately. Previous studies have focused on relatively large neocortical and subcortical samples. This low resolution imaging and subsequent image analysis have limited the accuracy of tissue sampling. Consequently, in the interest of obtaining pure tissue samples, areas potentially contributing to partially volumed CSF or metabolically active structures are typically avoided and at times corrections are estimated based on the corresponding CT or MRI scans.

Despite the limitations in spatial resolution, there is extensive literature describing the diagnostic utility of PET in studies of AD (Ferris et al. 1980; Benson et al. 1983; de Leon et al. 1983; Foster et al. 1984; Cutler et al. 1985; Friedland et al. 1985; Duara et al. 1986; Haxby et al. 1986). There is a consensus that temporal lobe and temporo-parietal reductions in glucose utilization are typically the most salient changes among the global changes seen in AD. Unfortunately, except for a few studies (Frackowiak et al. 1981; Duara et al. 1986; Jagust et al. 1988) there is very little longitudinal data published and consequently little is known about progressive patterns of brain change or early features or predictors of dementia. Moreover, there are no reports addressing hippocampal metabolic changes using current generation high resolution PET (i.e., less than 8 mm resolution in all planes). Our own work is in preparation.

To assess whether the structural hippocampal lesion is observed before the neocortical temporal lobe metabolic reduction, the metabolic rates for the normal controls and questionable dementia groups with and without hippocampal atrophy were compared. In this third study, hippocampal atrophy differentiated the groups as it was found in 1 of 21 normal and 7 of 9 questionable patients. Temporal lobe glucose metabolism and ventricular enlargement (not

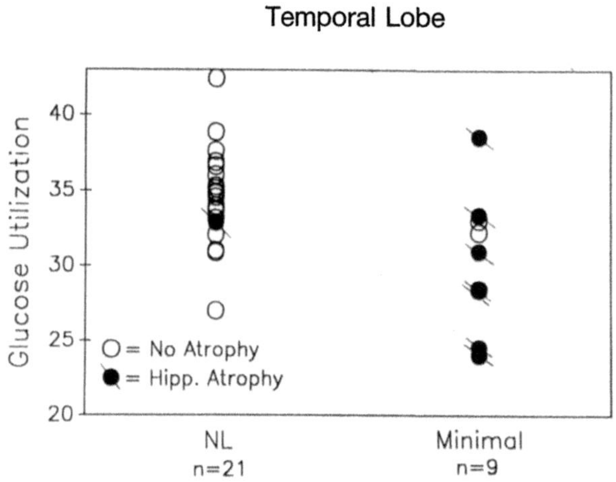

Fig. 5. The comparison of the PET glucose utilization study of the temporal lobe and the CT study of the hippocampal region in the differentiation of normal elderly from minimally impaired (non-demented) elderly patients

shown) did not reveal any group differentiation (Fig. 5). Therefore, with respect to the predictors of change, our longitudinal data show that CT hippocampal atrophy precedes PET temporo-parietal reductions in the questionably affected patients. The observations that structural hippocampal change precedes neocortical involvement further suggest that hippocampal involvement occurs earlier in the course of AD. The use of the high resolution PET will enable future studies to examine directly whether metabolic change in the hippocampus precedes structural hippocampal change.

MRI Studies

Using MRI, our objective has been to validate the anatomic and pathologic significance of the observed dilated choroidal fissures. In the *first validation study* we used contiguous 4 mm thick T1 weighted coronal MRI slices perpendicular to the long axis of the hippocampus to determine the volume of CSF in the region of the choroidal and hippocampal fissures. In all studies, the CSF volume was determined in three contiguous slices. The most anterior slice was just posterior to the Pes hippocampus. The results for 13 AD patients, five questionable AD subjects and eight age-matched elderly controls indicated that CSF volume, derived using an individualized threshold procedure, was significantly associated with the subjective ratings of dilated transverse fissures derived from 5 mm negative angulation transaxial T1 studies ($r = .68, p < .001$). In the region of maximum CSF accumulation (anterior), the MRI determined volume of CSF was increased in AD by 87% ($p < .04$, 1-tail). In the *second validation study* the same MRI protocol was used to derive the volume of several structures, including the hippocampus, to assess if there were volumetric reductions in the hippocampus associated with AD and if these reductions were specific anatomically. The structures investigated included the hippocampus, the parahippocampal gyrus, the fusiform gyrus, the cerebellum and the lateral temporal neocortex. As shown in Fig. 6, the largest volume losses for the three

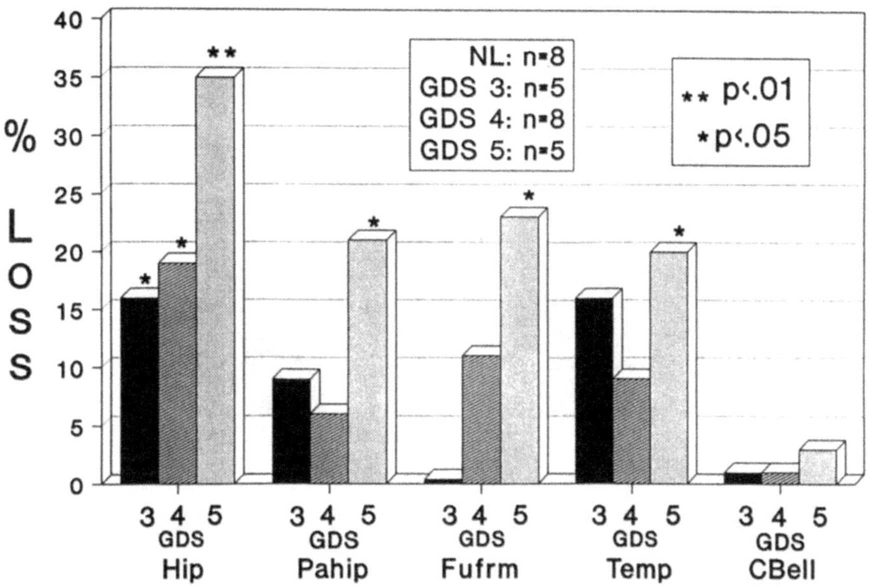

Fig. 6. Regional parenchymal MRI volume changes relative to control as a function of level of cognitive change (GDS)

patient groups (n' s = 5, 8 and 5) compared with control (n = 8) occur in the hippocampal formation. Moreover, the data show that across severity levels, the questionable or GDS 3 subjects with mild memory changes, but who did not meet criteria for AD, also show hippocampal volume reductions. These data therefore, lend support to the predictive validity of hippocampal atrophy in incipient AD cases. With respect to specificity of anatomical change, only the hippocampal measures show early significant reductions. Importantly these reduction are found in both the questionable (GDS 3) cases and in the mild AD cases (GDS 4).

In the *third validation study* we used postmortem MRI scans to determine the volume changes in the region of the hippocampus and parahippocampal gyrus. In this project four contiguous 3 mm T1 coronal images were studied on six confirmed AD cases and three age-matched normal controls. The normal and AD groups were 76.0 ± 3.6 and 80.8 ± 9.7 years of age, respectively. For comparison purposes, Fig. 7 shows the same anatomy studied in vivo for 13 AD patients and eight controls and with the volumes derived from the postmortem normal and AD cases. The postmortem results showed a 40% reduction of the parenchymal volume in the AD group. In the in vivo AD sample the volume reductions were approximately 20%. The difference in magnitude of change in the AD cases may be explained by greater patient severity in the postmortem compared to the in vivo cases. Note the comparable volumes for the two normal groups. Such results provide preliminary evidence for the suitability of making

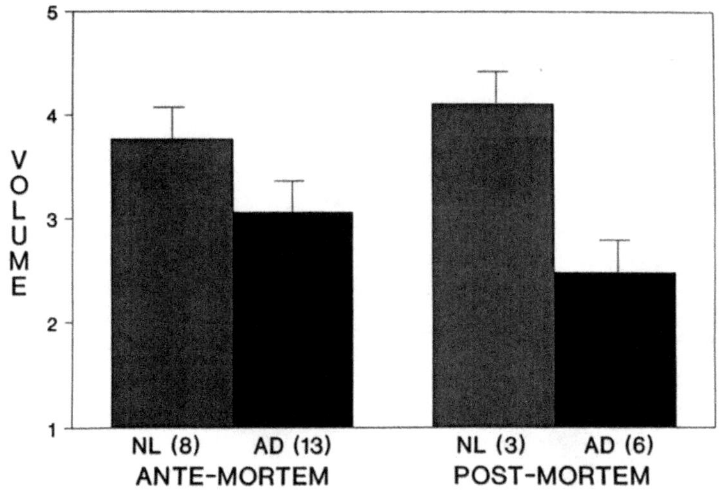

Fig. 7. Antemortem and postmortem MRI volumes of the parahippocampal gyrus including the hippocampal formation. Studies were obtained from four T1 weighted 3 mm coronal images

such measurements postmortem. Further, such measurements will provide the opportunity for precise MR correlation with neuropathology. These data also suggest the progressive nature of the hippocampal pathology throughout the course of the disease.

Postmortem Studies

On the six confirmed AD cases, we directly compared the postmortem hippocampal examination with both the antemortem CT and the postmortem MRI. The neuropathology protocol included coronal hippocampal samples taken at the level of the lateral geniculate nucleus and at other levels ($n = 5$). Following standard processing with paraffin, the formalin-fixed sections were stained with Luxol Fast Blue and counterstained with H and E. Congo Red and Bodian stains and 4G8 antibeta amyloid protein immunostain were also used. The AD patients showed extensive neurofibrillary tangles, amyloid deposition and neuronal loss in the hippocampal formation and parahippocampal gyrus when compared to the controls. All AD cases demonstrated neuropathologic and neuroimaging evidence of dilated transverse (choroidal-hippocampal) fissures.

Conclusions

In vivo examination of the hippocampal region appears to be of potential diagnostic and predictive value in the clinical study of AD. Hippocampal changes appear early in the natural history of AD and, when present in the absence of large neocortical metabolic brain changes, appear to be associated with the cognitive deficits so often ascribed to the normal aging process. Early hippocampal atrophy appears to be related to the AD process pathologically although it should be noted that there are no reports of autopsies on early AD patients who died of other causes. Nevertheless, it is universally observed that hippocampal atrophy occurs in AD, and we extend these findings to show that the in vivo imaging changes are directly related to degenerative parenchymal changes in the hippocampus.

Our longitudinal data indicate that hippocampal atrophy is predictive of the clinical dementia associated with AD. These findings now permit for the first time the identification of patients at risk for AD and therefore provide an early marker for potentially suitable drug study patients. Hippocampal atrophy also provides an opportunity to investigate the emergence of pathophysiologic phenomena and to evaluate their clinical significance. In particular, neuroendocrine disturbances are related to hippocampal degeneration in both animal models (Sapolsky et al. 1986) and in AD (de Leon et al. 1988). The extent to which these changes are related to the hippocampal or subsequent neocortical damage remains unknown. However, we have renewed optimism that with earlier diagnosis the search for relevant mechanisms of disease and effective treatment will be facilitated.

References

Ball MJ (1978) Topographic distribution of neurofibrillary tangles and granulovacuolar degeneration in hippocampal cortex of aging and demented patients. A quantitative study. Neuropathologica 42: 73–80

Ball MJ, Hachinski V, Fox A, Kirshen AJ, Fisman M, Blume W, Kral VA, Fox H (1985) A new definition of Alzheimer's disease: A hippocampal dementia. Lancet 1: 14–16

Benson DF, Kuhl DE, Hawkins RA, Phelps ME, Cummings JL, Tsai SY (1983) The fluorodeoxyglucose 18F scan in Alzheimer's disease and multi-infarct dementia. Arch Neurol 40: 711–714

Berg L, Danziger WL, Storandt M, Coben LA, Gado M, Hughes CP, Knesevich JW, Botwinick J (1984) Predictive features in mild senile dementia of the Alzheimer type. Neurology 34: 563–569

Cutler NR, Haxby JV, Duara R, Grady CL, Kay AD, Kessler RM, Sundaram M, Rapoport SI (1985) Clinical history, brain metabolism, and neuropsychological function in Alzheimer's disease. Ann Neurol 18: 298–309

Davies P, Maloney AJF (1976) Selective loss of central cholinergic neurons in Alzheimer's disease. Lancet 2: 1403

de la Monte SM (1989) Quantitation of cerebral atrophy in preclinical and end-stage Alzheimer's disease. Ann Neurol 25: 450–459

de Leon MJ, Ferris SH, George AE, Christman DR, Fowler JS, Gentes C, Reisberg B, Gee B, Emmerich M, Yonekura Y, Brodie J, Kricheff II, Wolf AP (1983) Positron emission tomography studies of aging and Alzheimer's disease. Am J Neuroradiol 4: 568–571

de Leon MJ, George AE, Stylopoulos LA, Smith G, Miller DC (1989) Early marker for Alzheimer's disease: the atrophic hippocampus. Lancet 672–673

de Leon MJ, McRae T, Tsai JR, George AE, Marcus DL, Freedman M, Wolf AP, McEwen B (1988) Abnormal hypercortisolemic response in Alzheimer's disease linked to hippocampal atrophy. Lancet 2: 391–392

Duara R, Grady C, Haxby J, Sundaram M, Cutler NR, Heston L, Moore A, Schlageter N, Larson S, Rapoport SI (1986) Positron emission tomography in Alzheimer's disease. Neurology 36: 879–887
Ferris SH, de Leon MJ, Wolf AP, Farkas T, Christman DR, Reisberg B, Fowler JS, MacGregor R, Goldman A, George AE, Rampal S (1980) Positron emission tomography in the study of aging and senile dementia. Neurobiol Aging 1: 127–131
Flicker C, Ferris SH, Crook T, Bartus RT (1987) Implications of memory and language dysfunction in the naming deficit of senile dementia. Brain Language 31: 187–200
Flicker C, Ferris SH, Reisberg B (1991) Mild cognitive impairment in the elderly: Predictors of dementia. Neurology 41: 1006–1009
Foster NL, Chase TN, Mansi L, Brooks R, Fedio P, Patronas NJ, Di Chiro G (1984) Cortical abnormalities in Alzheimer's disease. Ann Neurol 16: 649–654.
Frackowiak RSJ, Pozzilli C, Legg NJ, DuBoulay GH, Marshall J, Lenzi GL, Jones T (1981) Regional cerebral oxygen supply and utilization in dementia – a clinical and physiological study with oxygen-15 and positron tomography. Brain 104: 753–778
Friedland RP, Budinger TF, Koss E, Ober BA (1985) Alzheimer's disease: Anterior-posterior and lateral hemisphere alterations in cortical glucose utilization. Neurosci Lett 53: 235–240
George AE, de Leon MJ, Stylopoulos LA, Miller J, Kluger A, Smith G, Miller DC (1990) CT diagnostic features of Alzheimer disease: Importance of the choroidal/hippocampal fissure complex. Am J Neuroradiol 11: 101–107
Geschwind N (1967) The varieties of naming errors. Cortex 3: 97–112
Haxby JV, Grady CI, Duara R, Schlageter N, Berg G, Rapoport SI (1986) Neocortical metabolic abnormalities precede nonmemory cognitive defects in early Alzheimer's-type dementia. Arch Neurol 43: 882–885
Hughes CP, Berg L, Danziger WL, Cohen LA, Martin RL (1982) A new clinical scale for the staging of dementia. Br J Psychiatr 140: 566–572
Hyman BT, Van Hoesen GW, Damasio AR, Barnes CL (1984) Alzheimer's disease: Cell-specific pathology isolates the hippocampal formation. Science 225: 1168–1170
Hyman BT, Van Hoesen GW, Damasio AR (1987) Alzheimer's disease: glutamate depletion in the hippocampal perforant pathway zone. Ann Neurol 22: 37–40
Hyman BT, Van Hoesen GW (1988) Sites of earliest Alzheimer-type pathological changes in entorhinal cortex and hippocampus visualized by Alz-50 immunocytochemistry. Soc Neurosci Abstr 14 (Part 2): 1084. (Abstract)
Jagust WJ, Friedland RP, Budinger TF, Koss E, Ober B (1988) Longitudinal studies of regional cerebral metabolism in Alzheimer's disease. Neurology 38: 909–912
Kemper T (1984) Neuroanatomical and neuropathological changes in normal aging and in dementia. In: Albert L (ed) Clinical neurology of aging. Oxford University Press, New York
Kesslak JP, Nalcioglu O, Cotman CW (1991) Quantification of magnetic resonances scans for hippocampal and parahippocampal atrophy in Alzheimer's disease. Neurology 41: 51–54
Mann DMA, Esiri MM (1988) The site of the earliest lesions of Alzheimer's disease. New Engl J Med 318: 789–790
Milner B (1970) Memory and the medial temporal regions of the brain. In: Pribram KH, Broadbent DE (eds) Biological bases of memory. Academic, New York
Neary D, Snowden JS, Bowen DM, Sims NR, Mann DMA, Benton JS, Northen B, Yates PO, Davison AN (1986) Neuropsychological syndromes in presenile dementia due to cerebral atrophy. J Neurol Neurosurg Psychiat 49: 163–174
Powers RE, Struble RG, Casanova MF, O'Connor DT, Kitt CA, Price DL (1988) Innervation of human hippocampus by noradrenergic systems: Normal anatomy and structural abnormalities in aging and Alzheimer's disease. Neuroscience 25: 401–417
Reisberg B, Ferris SH, de Leon MJ, Crook T (1982) The global deterioration scale for assessment of primary degenerative dementia. Am J Psychiat 139: 1136–1139
Rubin EH, Morris JC, Grant EA, Vendegna T (1989) Very mild senile dementia of the Alzheimer type. I. Clinical assessment. Arch Neurol 46: 379–382
Sapolsky R, Krey LC, McEwen BS (1986) The neuroendocrinology of stress and aging: The glucocorticoid cascade hypothesis. Endocrine Rev 7: 284–301

Seab JP, Jagust WS, Wong SFS, Roos MS, Reed BR, Budinger TF (1988) Quantitative NMR measurements of hippocampal atrophy in Alzheimer's disease. Magnetic Resonance 8: 200–228

Selkoe D, Kosik K (1983) Neurochemical changes with aging. In: Albert M (ed) The clinical neurology of aging. Oxford University Press, Oxford, pp 53–75

Sim A, Sussman I (1962) Alzheimer's disease: Its natural history and differential diagnosis. J Nervous Mental Dis 135: 489–499

Storandt M, Hill RD (1989) Very mild senile dementia of the Alzheimer type. II. Psychometric test performance. Arch Neurol 46: 383–386

Terry RD, Peck A, De Teresa R, Schechter K, Horoupian DS (1981) Some morphometric aspects of the brain in senile dementia of the Alzheimer type. Ann Neurol 10: 184–192

Van Hoesen GW, Hyman BT, Damasio AR (1991) Entorhinal cortex pathology in Alzheimer's disease. Hippocampus 1: 1–8

Zola-Morgan S, Squire L, Amaral D (1986) Human amnesia and the medial temporal region: enduring memory impairment following a bilateral lesion limited to the field CA1 of the hippocampus. J Neurosci 6: 2950–2967

Consciousness: Evidence of Four Components in Alzheimer's Disease

J. L. Signoret

I am a neurologist. I have seen patients and I am fascinated by the brain and by philosophy. Perhaps what I am discussing is folk neurophilosophy. I will not try to define consciousness because all readers known what it means. I do not intend to illustrate a philosophical idea, or construct a neuronal model.

My purpose is to consider consciousness as the convergence of several cerebral functional components. This discussion is based on clinical pathological features observed in patients with Alzheimer's disease or degenerative dementia.

Four Pathological Cases

In the following four pathological cases one can say that consciousness is altered.

1. A patient is unable to recognize a car in a picture. This disorder of recognition affects many other objects. One may conclude that this patient has lost perceptive consciousness of objects. *Consciousness implies the correct recognition of objects.*
2. Another patient is unable to make his car start because it has broken down. He gets out of the car, opens the trunk, takes out the spare tire and changes one of the tires, even though it is not flat. This patient has lost consciousness of situations and is unable to resolve the problem of the breakdown. *Consciousness implies not only recognition of objects but also the capacity to resolve problems concerning the objects.*

These two cases permit us to state a first proposition: *Consciousness is necessary for a subject to establish a relationship to reality.*

3. A patient and his wife visit some friends in a nearby village. Two hours later, when it is time to go back, the patient does not remember how they got there. He has forgotten that they used the car. Yet he knows how to drive the car, how to repair it following a breakdown, and can recognize his car. *Consciousness implies conserving information.*
4. A patient consulting his doctor stains his hands with black ink from his pen. "It's all right", he says. A few minutes later the doctor asks him why his hands are dirty. The patient acknowledges that his hands are stained, and says it is because he repaired his car that morning. This patient has stored some

information but has lost consciousness of the events surrounding this information. *Consciousness implies not only conserving information but also the capacity to integrate information correctly into one's personal history ("self history").*

These two latter cases permit us to state a second proposition: *Consciousness is necessary for a subject to establish a relationship to his self history* (Table 1).

Allow me to repeat: Consciousness is necessary for a subject to establish a relationship to reality and to his self history.

Now I would like to propose a schema based on Charcot's famous bell, designed to illustrate the functional architecture of language components (Charcot 1884). There are four components: auditory, visual, phonetic and graphic (Fig. 1).

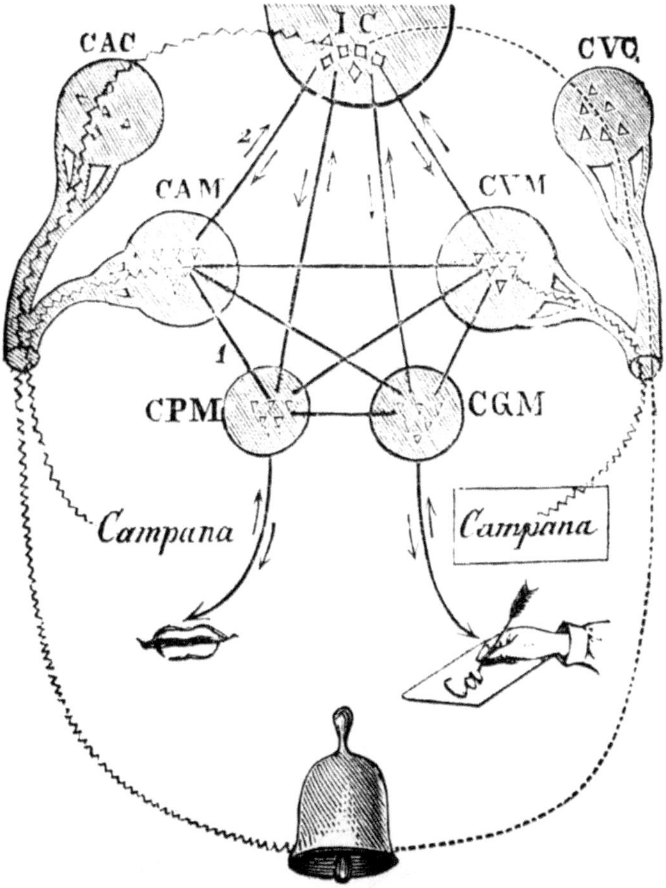

Fig. 1. The bell of Charcot. It illustrates the functional architecture of language with its four components: auditory (*CAM*), visual (*CVM*), phonetic (*CPM*), and graphic (*CGM*). (From Charcot 1884)

I am aware that I am a victim of my French history and *I know that reality can not and must not be only a car.*

The consciousness schema, with its four components – recognition, resolution, conservation, and integration – is shown in Fig. 2.

The important feature of this schema is that consciousness is not considered to be a superior faculty placed above the others. *Consciousness is the result of several components acting together at the same time, at all times.* There is an important paradox. Normally each of these components – recognition, resolution, conservation, integration – may function correctly in an unconscious

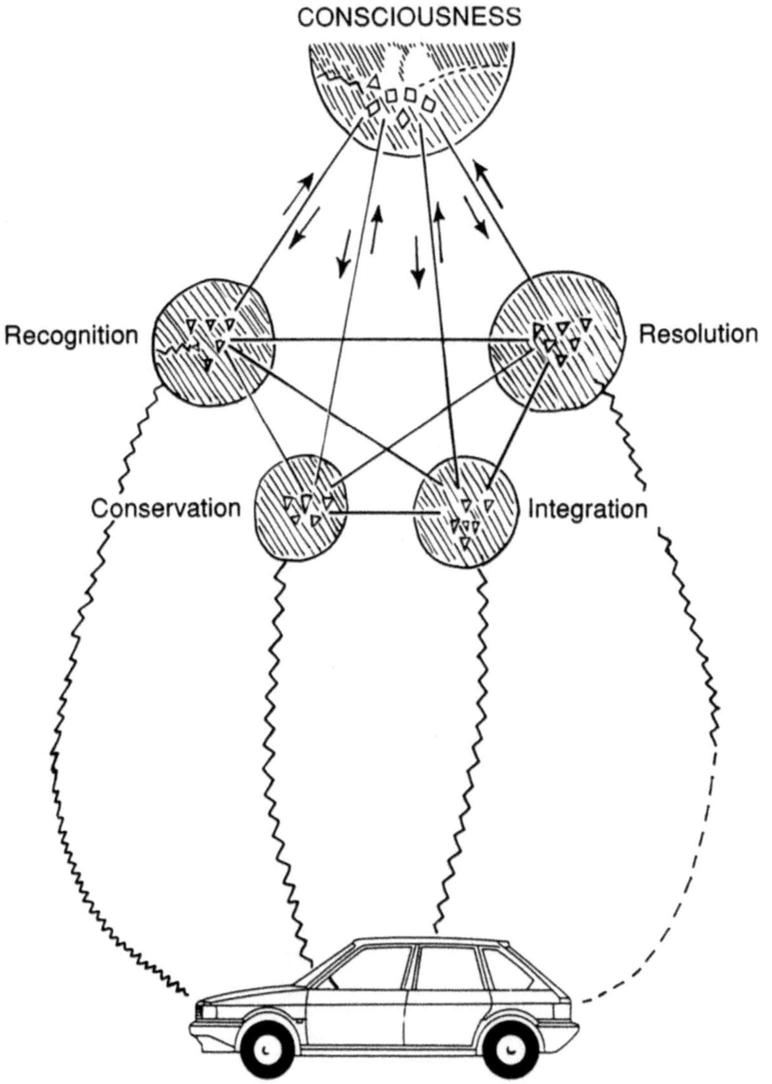

Fig. 2. The consciousness schema with its four components: *recognition, resolution, conservation* and *integration*

Table 1. The four components of consciousness

Case 1 A car is not recognized	Recognition
Case 2 A car breakdown is not resolved	Resolution
Case 3 A car journey is forgotten	Conservation
Case 4 An ink stain becomes a car stain	Integration

manner. *One can say that disorders of consciousness affect mental activities that may be normally unconscious.* "Unconscious" in this context is used for its cerebral meaning, not the psychoanalytic meaning.

I do not intend to propose neuronal models that underlie these components. I suggest that there are specific cerebral functional areas which underlie each of these four components. One can imagine a sort of map of cerebral topography which might be compatible with neuronal organization.

First Component: Recognition

Disorders of recognition can be the consequences of lesions of the associative cortex in the temporo-parieto-occipital areas. These disorders affect different function levels (Damasio 1985). One can also look at these disorders as lying between two extreme poles. At one pole the disorder affects the ability to perceive the visual fields as a whole, as, for instance, in Balint's syndrome, leading to unpredictable perception. At the other pole there is loss of recognition of objects; however, in some cases an unconscious recognition persists and allows the patient to classify objects according to categories (Damasio 1990).

Second Component: Resolution

Resolution of situations may be impaired following lesions of the dorsolateral frontal cortex. Situations which demand adjustment are those for which the patient has no immediate automatic response in his repertory of actions. In such situations he must plan out a strategy of action based on the representation of the situation, which implies correct recognition of the objects. The execution of this action must be controlled and checked (Shallice 1988). Depending on the difficulty of the situation the patient is confronted with, there are different functional levels of response. At one pole the disorder affects the normally automatic and unconscious adjustment. At the other pole there is loss of attention and thinking. At this level one must address the relationship between consciousness and free will.

There is also a social aspect included in situations, and a patient who ignores social rules in his conduct may exhibit a severe lack of adaptation. Such disorders are due to lesions of the ventro-medial frontal cortex (Eslinger and Damasio 1985). Other pathological features, such as utilization behavior, appear in cases of frontal lesions (Lhermitte 1983).

Third Component: Conservation

Disorders of conservation causing memory loss are apparent in cases of bilateral lesions of the ventromedial temporal lobe, especially in lesions affecting the hippocampus (Signoret 1985). Information that has been processed must be consolidated to be stored and recalled. I do not known what mechanisms are concerned with consolidation, but one can reasonably state that the hippocampal formation has a major role in the consolidation process which takes place in the synapses of the cortex. Depending on the extent of consolidation, there are different functional levels of memories. At one pole perfect and ideal consolidation leads to permanent memories that can be recalled at any moment. At the other pole a disorder of consolidation leads to fragile memories with incomplete recall. This incomplete recall can be the cause of partial consciousness. Is it a loss of consciousness when events that cannot be recalled may still be recognized?

Certain kind of memories do not need consolidation by the hippocampus because patients with Alzheimer's disease are still able to learn visuomotor tasks, and exhibit the effects of priming (Schachter 1987). These memories are unconscious. *It is interesting to note that consciousness depends on memory but memory does not always depends on consciousness.*

Fourth Component: Integration

Disorders of integration are due to lesions or dysfunction of the limbic system – the cingular cortex (Whitty and Lewin 1960) and the entorhinal cortex (Damasio et al. 1985); these two areas are interconnected and connected with prefrontal lobe.

All events can be memorized as long as they obey what one may call the principles of integration. An event occurs in a context, with reference to time and space, and all these elements are memorized together. At one pole a disorder of integration leads to source amnesia (Janowsky et al. 1989). Some information is stored but without good references to chronology, to time (Milner and Petrides 1984). Integration organizes information with relationship to time and space, but also according to other events already memorized. This permanent memorization of successive events of personal history is the basis of autobiographical memory (Rubin 1986). At the other pole, defective integration can mix up parts of recent events and parts of old events, leading to total chaos in time. Reference to time is lost, and patients may believe that past events are taking place in the present. Wernicke called this effect "allopsychic disorientation"; it is peculiar to a form of senile dementia, presbyophrenia (Berrios 1986).

However, such a disorder of integration may also affect the recall of memories, which cannot be correctly done without chronological order. Recall of memories is also concerned with integration.

Discussion

This discussion is probably incomplete because I have decided not to address two other problems:

- How the patient gains access to his mental activity, which leads to the relationship betwwen language and consciousness (Weiskrantz 1988).
- How disorders of personality affect consciousness, which leads to the relationship between emotional balance and consciousness.

I did not intend to break up consciousness into separate parts. Consciousness is not a belief or a desire. Yet consciousness is impossible to see. I have tried to analyse the complexity of consciousness by studying its components. From this study one can take either a reductionist or a constructivist position.

Figure 3 shows boxes that are never totally independent of each other. These boxes are built up to converge towards consciousness.

Reality and history – self history – play an important part and allow me to use clinical cases to justify my point of view. But there are two other possible boxes that can also lead to consciousness.

First, Alzheimer's disease patients may forget the identity of objects, people and sometimes even their own faces. The ability to identify an object is to be able to link the object with specific information. This information must be stored and integrated, so that an object can have an identity. This means there must be anatomic connections between the associative temporo-occipital areas and the

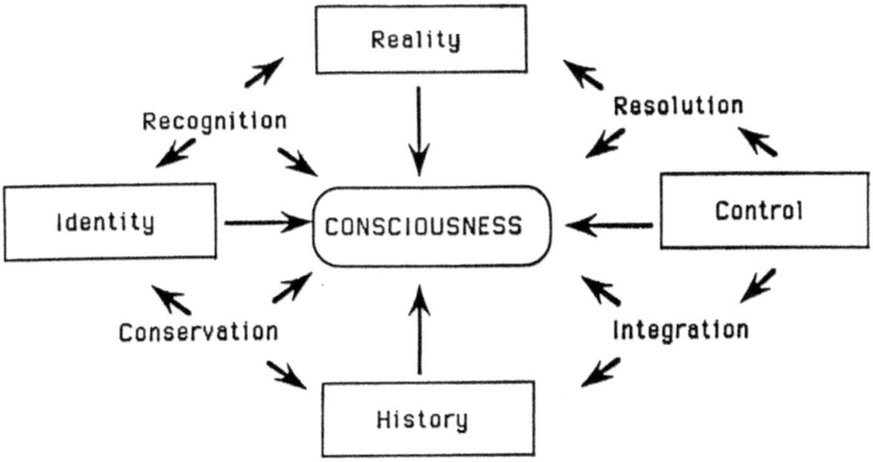

Fig. 3. The components which converge towards consciousness

ventro medial temporal lobes and the limbic system (Damasio 1989; Mesulam 1990).

Second, in Alzheimer's disease, disorders may affect all of the patient's initiatives which seem uncontrolled. Such control is necessary for effective adaptation to a situation, as it is for correct integration of the elements of an event. This means there must be anatomic connections between different parts of the frontal lobe, associative and limbic (Goldman-Rakic 1984).

I had proposed four components of consciousness. Are they recognition, resolution, conservation, and integration?

Are they reality, history, identify, and control?

Acknowledgements. I would like to thank Mlle Leslie Carts for her assistance in the English preparation of the manuscript.

References

Berrios GE (1986) Presbyophrenia: The rise and fall of a concept. Psychol Med 16: 267–275

Charcot JM (1884) Differenti forme d'afasia. Vallardi, Milan

Damasio AR (1985) Disorders of complex visual processing: agnosias, a-chromatopsia, Balint's syndrome, and related difficulties of orientation and construction. In: Mesulam MM (ed) Principles of behavioral neurology. Davis, Philadelphia, pp 259–288

Damasio AR (1989) Time-locked multiregional retroactivation: a systems-level proposal for the neural substrates of recall and recognition. Cognition 33: 25–62

Damasio AR (1990) Category-related recognition defects as a clue to the neural substrates of knowledge. Trends Neurosci 13: 95–98

Damasio AR, Eslinger P, Damasio H, Van Hoesen GW, Cornell S (1985) Multimodal amnesia syndrome following bilateral temporal and basal forebrain damage. Arch Neurol 42: 252–259

Eslinger P, Damasio A (1985) Severe disturbance of higher cognition after bilateral frontal lobe ablation. Neurology 35: 1731–1741

Goldman-Rakic PS (1984) The frontal lobes: uncharted provinces of the brain. Trends Neurosci 7: 425–429

Janowsky JS, Shimamura AP, Squire LR (1989) Source memory impairment in patients with frontal lobe lesions. Neuropsychologia 27: 1043–1056

Lhermitte F (1983) "Utilization behaviour" and its relation to lesions of the frontal lobes. Brain 106: 237–255

Mesulam MM (1990) Large-scale neurocognitive networks and distributed processing for attention, language and memory. Ann Neurol 28: 597–616

Milner B, Petrides M (1984) Behavioral effects of frontal lobe lesions in man. Trends Neurosci 7: 403–407

Rubin DC (1986) Autobiographical memory. Cambridge University Press, Cambridge

Schacter DL (1987) Implicit expressions of memory in organic amnesia: learning of new facts and associations. Hum Neurobiol 6: 107–118

Shallice T (1988) From neuropsychology to mental structure. Cambridge University Press, Cambridge

Signoret JL (1985) Memory and amnesias. In: Mesulam MM (ed) Principles of behavioral neuroly. Davis, Philadelphia, pp 169–192

Weiskrantz L (ed) (1988) Thought without language. Clarendon, Oxford

Whitty CNM, Lewin W (1960) A Korsakoff syndrome in the post cingulectomy confusional state. Brain 83: 648–653

Subject Index